# Differential Equations
## Workbook

FOR

# DUMMIES®

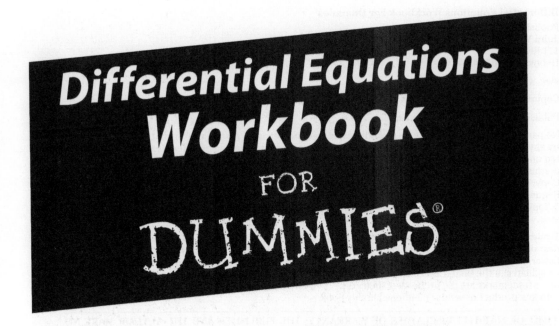

# Differential Equations Workbook FOR DUMMIES®

## by Steven Holzner, PhD

WILEY

John Wiley & Sons, Inc.

**Differential Equations Workbook For Dummies®**

Published by
**John Wiley & Sons, Inc.**
111 River St.
Hoboken, NJ 07030-5774

www.wiley.com

WILEY

# About the Author

**Steven Holzner** is the award-winning author of many books, including *Differential Equations For Dummies* and *Physics For Dummies*. He did his undergraduate work at MIT and got his PhD at Cornell University. He's been on the faculty of both MIT and Cornell.

# Dedication

To Nancy, of course.

# Author's Acknowledgments

Many people are responsible for putting this book together. My special thanks go out to Tracy Boggier, Chrissy Guthrie, Jen Tebbe, and technical editor Jamie Song, PhD. I'd also like to thank all the folks in Composition Services for their hard work creating equations and laying out the book.

## Publisher's Acknowledgments

We're proud of this book; please send us your comments through our Dummies online registration form located at http://dummies.custhelp.com. For other comments, please contact our Customer Care Department within the U.S. at 877-762-2974, outside the U.S. at 317-572-3993, or fax 317-572-4002.

Some of the people who helped bring this book to market include the following:

*Acquisitions, Editorial, and Media Development*

**Senior Project Editor:** Christina Guthrie

**Acquisitions Editor:** Tracy Boggier

**Copy Editor:** Jennifer Tebbe

**Assistant Editor:** Erin Calligan Mooney

**Editorial Program Coordinator:** Joe Niesen

**Technical Editor:** Jamie Song, PhD

**Editorial Manager:** Christine Meloy Beck

**Editorial Assistant:** David Lutton

**Cover Photos:** Kenneth Edward/
Photo Researchers, Inc.

**Cartoons:** Rich Tennant (www.the5thwave.com)

*Composition Services*

**Project Coordinator:** Lynsey Stanford

**Layout and Graphics:** Carrie A. Cesavice,
Reuben W. Davis, Mark Pinto, Christine Williams

**Proofreaders:** Laura Albert, Cynthia Fields

**Indexer:** Ty Koontz

---

**Publishing and Editorial for Consumer Dummies**

    **Kathleen Nebenhaus,** Vice President and Executive Publisher

    **David Palmer,** Associate Publisher

    **Kristin Ferguson-Wagstaffe,** Product Development Director

**Publishing for Technology Dummies**

    **Andy Cummings,** Vice President and Publisher

**Composition Services**

    **Debbie Stailey,** Director of Composition Services

# Contents at a Glance

# Table of Contents

# Introduction

*T*oo often, differential equations seem like torture. They seem so bad in fact that you may be tempted to cringe or shudder when you're assigned homework that involves 'em.

*Differential Equations Workbook For Dummies* may not get you to embrace differential equations with open arms, but it *will* improve your understanding of the pesky things. Here you get ample practice working through the most common types of differential equations, along with detailed solutions, so you can truly master the subject. Get ready to add "differential equations expert" to your résumé!

## About This Book

*Differential Equations Workbook For Dummies* is all about practicing solving differential equations. It's crammed full of the good stuff — and *only* the good stuff. Each aspect of differential equations is addressed with some brief text to refresh your memory of the basics, a worked-out example, and multiple practice problems. (If you're looking for in-depth explanation of differential equations topics, your best resource is *Differential Equations For Dummies* [Wiley] or your class textbook.) So that you're not left hanging wondering whether your solution is right or wrong, each chapter features an answers section with all the practice problems worked out, step by glorious step.

You can leaf through this workbook as you like, solving problems and reading solutions as you go. Like other *For Dummies* books, this one is designed to let you skip around to your heart's content.

## Conventions Used in This Book

Some books have a dozen confusing conventions that you need to know before you can even start. Not this one. You need to keep just these few things in mind:

- New terms appear in *italics* the first time they're presented. And like other math books, this one also employs italics to indicate variables.

- Web sites appear in monofont to help them stand out. (In some cases, a Web site may break across multiple lines. Rest assured I haven't inserted any extra spaces or punctuation; just type the address as provided.)

- In the answers section at the end of every chapter, the practice problems and solutions appear in **bold** (the step-by-step info that follows is in regular text). Matrices and keywords in bulleted lists are also given in bold.

# Foolish Assumptions

Any study of differential equations takes knowledge of calculus as its starting point. You should know how to take basic derivatives and how to integrate before reading this workbook (and if you don't, I recommend picking up a copy of *Calculus For Dummies* [Wiley] first).

Most importantly, I'm assuming you already have an in-depth resource about differential equations available to you. This workbook is intended to give you extra practice tackling standard differential equations concepts; it doesn't provide detailed instruction on the fundamentals of differential equations. I do include some brief refresher text on each aspect of differential equations, but if you're brand-new to the subject, check out *Differential Equations For Dummies* or your class textbook.

# How This Book Is Organized

This workbook is organized modularly, into parts, following the same organizational structure as *Differential Equations For Dummies*. Here's what you're going to find in each part.

## Part 1: Tackling First Order Differential Equations

First order differential equations are the easiest differential equations to solve. That's why this part gives you practice finding solutions to linear, separable, and exact first order differential equations.

## Part II: Finding Solutions to Second and Higher Order Differential Equations

The most interesting differential equations used in the real world are second order differential equations. Here, you practice multiple ways of solving this type of equation. You also get to try your hand at solving third and higher order differential equations. Things get pretty steep pretty fast, but fortunately you have some surprising techniques at your disposal, as you discover in this part.

## Part III: The Power Stuff: Advanced Techniques

I've pulled out all the stops in Part III. In these chapters, you find some powerful solution techniques, including power series, which you can use to convert a tough differential equation into an algebra problem and then solve for the coefficients of each power, and Laplace transforms, which can occasionally give you the solution you're looking for in no time.

## Part IV: The Part of Tens

The classic *For Dummies* Part of Tens provides you with a couple collections of top ten resources. Flip to this part to find help with the ten common ways of solving differential equations or to discover ten real-world applications for differential equations.

# Icons Used in This Book

*For Dummies* books always use icons to point out important information; this workbook is no different. Here's the quick-and-dirty of what the icons mean:

This icon points out practice problems that have been worked out for you to get you off on the right foot.

Looking for the juicy tidbits that are essential to your study of differential equations? Then watch for paragraphs marked with this icon.

This icon denotes tricks and techniques to make your life easier (at least as it relates to solving differential equations).

# Where to Go from Here

You can start anywhere you feel you need the most practice. In fact, this workbook was written to allow you to do just that. However, if you want to follow along with *Differential Equations For Dummies* or your textbook, your best bet is to start with Chapter 1.

You may also want to grab a few pieces of scratch paper. I've tried to leave you enough room to work the problems right in the book, but you still might find a little extra paper helpful.

# Part I

# Tackling First Order Differential Equations

The 5th Wave          By Rich Tennant

MATH MEN'S CLUB

Let's hear it for Amber, who's gonna show us what she knows abooooUut first order differential equaaations!

## In this part . . .

Welcome to the world of first order differential equations! Here, you put your skills to the test with linear first order differential equations, which means you're dealing with first order derivatives that are to the first power, not the second or any other higher power. You also work with separable first order differential equations, which can be separated so that only terms in $y$ appear on one side of the equation and only terms in $x$ appear on the other side (okay, okay, constants can appear on this side too). Finally, you practice solving exact differential equations.

# Chapter 1

# Looking Closely at Linear First Order Differential Equations

............................................................

## In This Chapter

▶ Knowing what a first order linear differential equation looks like

▶ Finding solutions to first order differential equations with and without *y* terms

▶ Employing the trick of integrating factors

............................................................

**O**ne important way that you can classify differential equations is as linear or nonlinear. A differential equation is considered *linear* if it involves only *linear terms* (that is, terms to the power 1) of *y*, *y′*, *y″*, and so on. The following equation is an example of a linear differential equation:

$$L\frac{d^2Q}{dx^2} + R\frac{dQ}{dx} + \frac{1Q}{C} = E(x)$$

*Nonlinear* differential equations simply include nonlinear terms in *y*, *y′*, *y″*, and so on. This next equation, which describes the angle of a pendulum, is considered a nonlinear differential equation because it involves the term sin θ (not just θ):

$$\frac{d^2\theta}{dx^2} + \frac{g}{L}\sin\theta = 0$$

This chapter focuses on linear first order differential equations. Here you have the chance to sharpen your linear-equation-spotting eye. You also get to practice solving linear first order differential equations when *y* is and isn't involved. Finally, I clue you in to a little (yet extremely useful!) trick o' the trade called integrating factors.

## Identifying Linear First Order Differential Equations

Here's the general form of a linear differential equation, where $p(x)$ and $q(x)$ are functions (which can just be constants):

$$\frac{dy}{dx} + p(x)y = q(x)$$

Following are some examples of linear differential equations:

$$\frac{dy}{dx} = 5$$

$$\frac{dy}{dx} = y + 1$$

$$\frac{dy}{dx} = 3y + 1$$

For a little practice, try to figure out whether each of the following equations is linear or nonlinear.

**Q.** Is this equation a linear first order differential equation?

$$\frac{dy}{dx} = 17y + 4$$

**A.** Yes.

This equation is a linear first order differential equation because it involves solely first order terms in $y$ and $y'$.

**1.** Is this equation a linear first order differential equation?

$$\frac{dy}{dx} = 9y + 1$$

Solve It

**2.** Is the following a linear first order differential equation?

$$\frac{dy}{dx} = 17y^3 + 4$$

Solve It

**3.** Is this equation a linear first order differential equation?

$$\frac{dy}{dx} = y\cos(x)$$

Solve It

**4.** Is the following a linear first order differential equation?

$$\frac{dy}{dx} = x\cos(y)$$

Solve It

# Solving Linear First Order Differential Equations That Don't Involve Terms in y

The simplest type of linear first order differential equation doesn't have a term in $y$ at all; instead, it involves just the first derivative of $y$, $y'$, $y''$, and so on. These differential equations are simple to solve because the first derivatives are easy to integrate. Here's the general form of such equations (note that $q(x)$ is a function, which may be a constant):

$$\frac{dy}{dx} = q(x)$$

Take a look at this linear first order differential equation:

$$\frac{dy}{dx} = 3$$

Note that there's no term in just $y$. So how do you solve this kind of equation? Just move the $dx$ over to the right:

$$dy = 3dx$$

Then integrate to get

$$y = 3x + c$$

where $c$ is a constant of integration.

To figure out what $c$ is, simply take a look at the initial conditions. For example, say that $y(0)$ — that is, the value of $y$ when $x = 0$ — is equal to

$$y(0) = 15$$

Plugging $y(0) = 15$ into $y = 3x + c$ gives you

$$y(0) = c = 15$$

So $c = 15$ and $y = 3x + 15$. That's the complete solution!

To deal with constants of integration like $c$, look for the specified initial conditions. For example, the problem you just solved is usually presented as

$$\frac{dy}{dx} = 3$$

where

$$y(0) = 15$$

Time for a more advanced problem! (Note that this one still doesn't involve any simple terms in $y$.)

$$\frac{dy}{dx} = x^3 - 3x^2 + x$$

where

$$y(0) = 3$$

Because this equation doesn't involve any terms in $y$, you can move the $dx$ to the right, like this:

$$dy = x^3\, dx - 3x^2\, dx + x\, dx$$

Then just integrate to get

$$y = \frac{x^4}{4} - x^3 + \frac{x^2}{2} + c$$

To evaluate $c$, use the initial condition, which is

$$y(0) = 3$$

Plugging $x = 0 \rightarrow y = 3$ into the equation for $y$ gives you

$$y(0) = 3 = c$$

So the full solution is

$$y = \frac{x^4}{4} - x^3 + \frac{x^2}{2} + 3$$

As you can see, the way to deal with linear first order differential equations that don't involve a term in just $y$ is simply to

1. **Move the $dx$ to the right and integrate.**

2. **Apply the initial conditions to solve for the constant of integration.**

Following are some practice problems to make sure you have the hang of it.

**Q.** Solve for $y$ in this differential equation:

$$\frac{dy}{dx} = 2x$$

where

$$y(0) = 3$$

**A.** $y = x^2 + 3$

    1. Multiply both sides by $dx$:

$$dy = 2x \, dx$$

2. Integrate both sides to get the following, where $c$ is a constant of integration:

$$y = x^2 + c$$

3. Apply the initial condition to get

$$c = 3$$

4. Having solved for $c$, you can find the solution to the differential equation:

$$y = x^2 + 3$$

---

**5.** Solve for $y$ in this differential equation:

$$\frac{dy}{dx} = 8x$$

where

$$y(0) = 4$$

*Solve It*

**6.** What's $y$ in the following equation?

$$\frac{dy}{dx} = 2x + 2$$

where

$$y(0) = 2$$

*Solve It*

**7.** Solve for $y$ in this differential equation:

$$\frac{dy}{dx} = 6x + 5$$

where

$$y(0) = 10$$

*Solve It*

**8.** What's $y$ in the following equation?

$$\frac{dy}{dx} = 8x + 3$$

where

$$y(0) = 12$$

*Solve It*

# Solving Linear First Order Differential Equations That Involve Terms in y

Wondering what to do if a differential equation you're facing involves both $x$ and $y$?

$$\frac{dy}{dx} + p(x)\, y = q(x)$$

Start by taking a look at this representative problem:

$$\frac{dy}{dx} = ay - b$$

The preceding is a linear first order differential equation that contains both $dy/dx$ and $y$. How do you handle it and find a solution? By using some algebra, you can rewrite this equation as

$$\frac{dy/dx}{y - (b/a)} = a$$

Multiplying both sides by $dx$ gives you

$$\frac{dy}{y-(b/a)} = a \, dx$$

Congrats! You've just separated $x$ on one side of this differential equation and $y$ on the other, making the integration much easier. Speaking of integration, integrating both sides gives you

$$\ln |y - (b/a)| = ax + C$$

where $C$ is a constant of integration. Raising both sides to the power $e$ gives you this, where $c$ is a constant defined by $c = e^C$:

$$y = (b/a) + ce^{ax}$$

Anything beyond this level of difficulty must be approached in another way, and you deal with such equations throughout the rest of the book.

If you think you have solving linear first order differential equations in terms of $y$ all figured out, try your hand at these practice questions.

**Q.** Solve for $y$ in this differential equation:

$$\frac{dy}{dx} = 2y - 4$$

where

$$y(0) = 3$$

**A.** $y = 2 + e^{2x}$

1. Use algebra to get

$$\frac{dy/dx}{y-2} = 2$$

2. Then multiply both sides by $dx$:

$$\frac{dy}{y-2} = 2dx$$

3. Integrate to get

$$\ln |y - 2| = 2x + C$$

4. Then raise $e$ to the power of both sides:

$$y = 2 + e^C e^{2x} = 2 + ce^{2x}$$

5. Finally, apply the initial condition to get

$$y = 2 + e^{2x}$$

**9.** What's $y$ in the following equation?

$$\frac{dy}{dx} = 4y - 8$$

where

$$y(0) = 5$$

Solve It

**10.** Solve for $y$ in this differential equation:

$$\frac{dy}{dx} = 3y - 9$$

where

$$y(0) = 9$$

Solve It

**11.** What's $y$ in the following equation?

$$\frac{dy}{dx} = 9y - 18$$

where

$$y(0) = 5$$

Solve It

**12.** Solve for $y$ in this differential equation:

$$\frac{dy}{dx} = 4y - 20$$

where

$$y(0) = 16$$

Solve It

# Integrating Factors: A Trick of the Trade

Because not all differential equations are as nice and neat to work with as the ones featured earlier in this chapter, you need to have more power in your differential equation–solving arsenal. Enter *integrating factors,* which are functions of μ(x). The idea behind an integrating factor is to multiply the differential equation by it so that the resulting equation can be integrated easily.

Say you encounter this differential equation:

$$\frac{dy}{dx} + 3y = 9$$

where

$$y(0) = 7$$

To solve this equation with an integrating factor, try multiplying by μ(x), your as-yet-undetermined integrating factor:

$$\mu(x)\,\frac{dy}{dx} + 3\mu(x)y = 9\mu(x)$$

The trick now is to select μ(x) so you can recognize the left side as a derivative of something that can be easily integrated. If you take a closer look, you notice that the left side of this equation appears very much like differentiating the product μ(x)y, because the derivative of μ(x)y with respect to x is

$$\frac{d\big(\mu(x)y\big)}{dx} = \mu(x)\,\frac{dy}{dx} + y\,\frac{d\mu(x)}{dx}$$

Comparing the right side of this differential equation to the left side of the previous one gives you

$$\frac{d\mu(x)}{dx} = 3\mu(x)$$

At last! That looks like something you can work with. Rearrange the equation to get the following:

$$\frac{d\mu(x)/dx}{\mu(x)} = 3$$

Then go ahead and multiply both sides by dx to get

$$\frac{d\mu(x)}{\mu(x)} = 3dx$$

Integrating gives you

$$\ln|\mu(x)| = 3x + b$$

where b is a constant of integration.

Raising $e$ to the power of both sides gives you

$$\mu(x) = ce^{3x}$$

where $c$ is another constant ($c = e^b$).

Guess what? You've just found an integrating factor, specifically $\mu(x) = ce^{3t}$.

You can use that integrating factor with the original differential equation, multiplying the equation by $\mu(x)$:

$$\mu(x)\frac{dy}{dx} + 3\mu(x)y = 9\mu(x)$$

which is equal to

$$ce^{3x}\frac{dy}{dx} + 3ce^{3x}y = 9ce^{3x}$$

As you can see, the constant $c$ drops out, leaving you with

$$e^{3x}\frac{dy}{dx} + 3e^{3x}y = 9e^{3x}$$

TIP

Because you're only looking for a multiplicative integrating factor, you can either drop the constant of integration when you find an integrating factor or set $c = 1$.

This is where the whole genius of integrating factors comes in, because you can recognize the left side of this equation as the derivative of the product $e^{3x}y$. So the equation becomes

$$\frac{d(e^{3x}y)}{dx} = 9e^{3x}$$

That sure looks a lot easier to handle than the original version of this differential equation, doesn't it?

Now you can multiply both sides by $dx$ to get

$$d(e^{3x}y) = 9e^{3x}\,dx$$

Then integrate both sides:

$$e^{3x}y = 3e^{3x} + c$$

and solve for $y$:

$$y = 3 + ce^{-3x}$$

Because the initial condition stated that $y(0) = 7$, that means $c = 4$, so

$$y = 3 + 4e^{-3x}$$

Pretty cool, huh?

Here are some practice equations to get you better acquainted with the trick of integrating factors.

**Q.** Solve for $y$ by using an integrating factor:

$$\frac{dy}{dx} + 5y = 10$$

where

$$y(0) = 6$$

**A.** $y = 2 + 4e^{-5x}$

1. Multiply both sides of the differential equation by $\mu(x)$ to get

$$\mu(x)\frac{dy}{dx} + 5\mu(x)y = 10\mu(x)$$

2. Identify the left side with a derivative (in this case, the derivative of a product):

$$\frac{d(\mu(x)y)}{dx} = \mu(x)\frac{dy}{dx} + \frac{d\mu(x)}{dx}y$$

3. Then identify the right side of the equation in Step 2 with the left side of the equation in Step 1:

$$\frac{d\mu(x)}{dx} = 5\mu(x)$$

4. Rearrange terms to get

$$\frac{d\mu(x)}{\mu(x)} = 5dx$$

5. Then integrate:

$$\ln|\mu(x)| = 5x + b$$

6. Next up, raise $e$ to the power of both sides(where $c = e^b$) to get

$$\mu(x) = ce^{5x}$$

7. Multiply the original differential equation by the integrating factor (canceling out $c$) to get

$$e^{5x}\frac{dy}{dx} + 5e^{5x}y = 10e^{5x}$$

8. Combine the terms on the left side of this equation:

$$\frac{d(e^{5x}y)}{dx} = 10e^{5x}$$

9. Then multiply by $dx$:

$$d(e^{5x}y) = 10e^{5x}\,dx$$

10. Integrate:

$$e^{5x}y = 2e^{5x} + c$$

11. Divide both sides by $e^{5x}$ to get

$$y = 2 + ce^{-5x}$$

12. Finally, apply the initial condition:

$$y = 2 + 4e^{-5x}$$

**13.** Solve for $y$ by using an integrating factor:

$$\frac{dy}{dx} + 2y = 4$$

where

$$y(0) = 3$$

*Solve It*

**14.** In the following differential equation, find $y$ by using an integrating factor:

$$\frac{dy}{dx} + 3y = 9$$

where

$$y(0) = 8$$

*Solve It*

**15.** Solve for $y$ by using an integrating factor:

$$\frac{dy}{dx} + 2y = 14$$

where

$$y(0) = 9$$

*Solve It*

**16.** In the following differential equation, find $y$ by using an integrating factor:

$$\frac{dy}{dx} + 9y = 63$$

where

$$y(0) = 8$$

*Solve It*

# Answers to Linear First Order Differential Equation Problems

Following are the answers to the practice questions presented throughout this chapter. Each one is worked out step by step so that if you messed one up along the way, you can more easily see where you took a wrong turn.

**1** Is this equation a linear first order differential equation?

$$\frac{dy}{dx} = 9y + 1$$

**Yes.** This equation is a linear first order differential equation because it involves solely first order terms in $y$ and $y'$.

**2** Is the following a linear first order differential equation?

$$\frac{dy}{dx} = 17y^3 + 4$$

**No.** This equation is *not* a linear first order differential equation because it doesn't involve solely first order terms in $y$ and $y'$.

**3** Is this equation a linear first order differential equation?

$$\frac{dy}{dx} = y \cos(x)$$

**No.** This equation is *not* a linear first order differential equation because it doesn't involve solely first order terms in $y$ and $y'$.

**4** Is the following a linear first order differential equation?

$$\frac{dy}{dx} = x \cos(y)$$

**No.** This equation is *not* a linear first order differential equation because it doesn't involve solely first order terms in $y$ and $y'$.

**5** Solve for $y$ in this differential equation:

$$\frac{dy}{dx} = 8x$$

where

$$y(0) = 4$$

**Solution:** $y = 4x^2 + 4$

1. Multiply both sides by $dx$:

$$dy = 8x \, dx$$

2. Then integrate both sides to find the following, where $c$ is a constant of integration:

   $y = 4x^2 + c$

3. Apply the initial condition to get

   $c = 4$

4. Having solved for $c$, you can now find the solution, which is

   $y = 4x^2 + 4$

**6**   **What's $y$ in the following equation?**

   $$\frac{dy}{dx} = 2x + 2$$

   **where**

   $y(0) = 2$

**Solution: $y = x^2 + 2x + 2$**

1. Start by multiplying both sides by $dx$:

   $dy = 2x\, dx + 2dx$

2. Integrate both sides (noting that $c$ is a constant of integration):

   $y = x^2 + 2x + c$

3. Apply the initial condition:

   $c = 2$

4. Having solved for $c$, obtain the solution to the equation:

   $y = x^2 + 2x + 2$

**7**   **Solve for $y$ in this differential equation:**

   $$\frac{dy}{dx} = 6x + 5$$

   **where**

   $y(0) = 10$

**Solution: $y = 3x^2 + 5x + 10$**

1. Multiply both sides by $dx$:

   $dy = 6x\, dx + 5dx$

2. Integrate both sides to find the following, where $c$ is a constant of integration:

   $y = 3x^2 + 5x + c$

3. Apply the initial condition to get

   $c = 10$

4. Having solved for $c$, you can now find the solution, which is

   $y = 3x^2 + 5x + 10$

**8** **What's $y$ in the following equation?**

$$\frac{dy}{dx} = 8x + 3$$

where

$$y(0) = 12$$

**Solution: $y = 4x^2 + 3x + 12$**

1. Start by multiplying both sides by $dx$:

   $$dy = 8x\,dx + 3dx$$

2. Integrate both sides (noting that $c$ is a constant of integration):

   $$y = 4x^2 + 3x + c$$

3. Apply the initial condition:

   $$c = 12$$

4. Having solved for $c$, obtain the solution to the equation:

   $$y = 4x^2 + 3x + 12$$

**9** **What's $y$ in the following equation?**

$$\frac{dy}{dx} = 4y - 8$$

where

$$y(0) = 5$$

**Solution: $y = 2 + 3e^{4x}$**

1. First, use algebra to get

   $$\frac{dy/dx}{y-2} = 4$$

2. Then multiply both sides by $dx$:

   $$\frac{dy}{y-2} = 4dx$$

3. Integrate:

   $$\ln|y-2| = 4x + c$$

4. Raise $e$ to the power of both sides:

   $$y = 2 + ce^{4x}$$

5. Finally, apply the initial condition:

   $$y = 2 + 3e^{4x}$$

**10** Solve for *y* in this differential equation:

$$\frac{dy}{dx} = 3y - 9$$

where

$$y(0) = 9$$

**Solution:** $y = 3 + 6e^{3x}$

1. Use algebra to change the equation to

$$\frac{dy/dx}{y-3} = 3$$

2. Multiply both sides by *dx*:

$$\frac{dy}{y-3} = 3dx$$

3. Then integrate to get

$$\ln|y-3| = 3x + c$$

4. Raise *e* to the power of both sides:

$$y = 3 + ce^{3x}$$

5. Last but not least, apply the initial condition to get

$$y = 3 + 6e^{3x}$$

**11** What's *y* in the following equation?

$$\frac{dy}{dx} = 9y - 18$$

where

$$y(0) = 5$$

**Solution:** $y = 2 + 3e^{9x}$

1. First, use algebra to get

$$\frac{dy/dx}{y-2} = 9$$

2. Then multiply both sides by *dx*:

$$\frac{dy}{y-2} = 9dx$$

3. Integrate:

$$\ln|y-2| = 9x + c$$

4. Raise *e* to the power of both sides:

$$y = 2 + ce^{9x}$$

5. Finally, apply the initial condition:

$$y = 2 + 3e^{9x}$$

**12** Solve for $y$ in this differential equation:

$$\frac{dy}{dx} = 4y - 20$$

where

$$y(0) = 16$$

Solution: $y = 5 + 11e^{4x}$

1. Use algebra to change the equation to

$$\frac{dy/dx}{y - 5} = 4$$

2. Multiply both sides by $dx$:

$$\frac{dy}{y - 5} = 4dx$$

3. Then integrate to get

$$\ln|y - 5| = 4x + c$$

4. Raise $e$ to the power of both sides:

$$y = 5 + ce^{4x}$$

5. Last but not least, apply the initial condition to get

$$y = 5 + 11e^{4x}$$

**13** Solve for $y$ by using an integrating factor:

$$\frac{dy}{dx} + 2y = 4$$

where

$$y(0) = 3$$

Solution: $y = 2 + e^{-2x}$

1. Multiply both sides of the equation by $\mu(x)$ to get

$$\mu(x)\frac{dy}{dx} + 2\mu(x)y = 4\mu(x)$$

2. Identify the left side with a derivative (in this case, the derivative of a product):

$$\frac{d(\mu(x)y)}{dx} = \mu(x)\frac{dy}{dx} + \frac{d\mu(x)}{dx}y$$

3. Then identify the right side of the equation in Step 2 with the left side of the equation in Step 1:

$$\frac{d\mu(x)}{dx} = 2\mu(x)$$

4. Rearrange the terms to get

$$\frac{d\mu(x)}{\mu(x)} = 2dx$$

5. Then integrate:

   $\ln |\mu(x)| = 2x + b$

6. Raise $e$ to the power of both sides (where $c = e^b$) to get

   $\mu(x) = ce^{2x}$

7. Multiply the original differential equation by the integrating factor (canceling out $c$):

   $e^{2x} \dfrac{dy}{dx} + 2e^{2x}y = 4e^{2x}$

8. Then combine the terms on the left side of this equation to get

   $\dfrac{d\left(e^{2x}y\right)}{dx} = 4e^{2x}$

9. Next, multiply by $dx$:

   $d(e^{2x}y) = 4e^{2x}\,dx$

10. Integrate:

    $e^{2x}y = 2e^{2x} + c$

11. Then divide both sides by $e^{2x}$ to get

    $y = 2 + ce^{-2x}$

12. Finally, apply the initial condition to achieve your answer:

    $y = 2 + e^{-2x}$

**14** In the following differential equation, find $y$ by using an integrating factor:

$\dfrac{dy}{dx} + 3y = 9$

**where**

$y(0) = 8$

**Solution: $y = 3 + 5e^{-3x}$**

1. Multiply both sides by $\mu(x)$:

   $\mu(x)\dfrac{dy}{dx} + 3\mu(x)y = 9\mu(x)$

2. Identify the left side of the equation with a derivative (in this case, the derivative of a product):

   $\dfrac{d\left(\mu(x)y\right)}{dx} = \mu(x)\dfrac{dy}{dx} + \dfrac{d\mu(x)}{dx}y$

3. Identify the right side of the equation in Step 2 with the left side of the equation in Step 1:

   $\dfrac{d\mu(x)}{dx} = 3\mu(x)$

4. Next, rearrange the terms:

$$\frac{d\mu(x)}{\mu(x)} = 3dx$$

5. Integrate to get

$$\ln |\mu(x)| = 3x + b$$

6. Raise $e$ to the power of both sides (where $c = e^b$):

$$\mu(x) = ce^{3x}$$

7. Multiply the original equation by the integrating factor (canceling out $c$) to get

$$e^{3x}\frac{dy}{dx} + 3e^{3x}y = 9e^{3x}$$

8. Combine terms on the left side of the equation:

$$\frac{d\left(e^{3x}y\right)}{dx} = 3e^{3x}$$

9. Multiply by $dx$:

$$d(e^{3x}y) = 3e^{3x}\,dx$$

10. Then integrate to get

$$e^{3x}y = 3e^{3x} + c$$

11. Next, divide both sides by $e^{3x}$:

$$y = 3 + ce^{-3x}$$

12. After applying the initial condition, you should have

$$y = 3 + 5e^{-3x}$$

---

**15** **Solve for $y$ by using an integrating factor:**

$$\frac{dy}{dx} + 2y = 14$$

**where**

$$y(0) = 9$$

**Solution: $y = 7 + 2e^{-2x}$**

1. Multiply both sides of the equation by $\mu(x)$ to get

$$\mu(x)\frac{dy}{dx} + 2\mu(x)y = 14\mu(x)$$

2. Identify the left side with a derivative (in this case, the derivative of a product):

$$\frac{d\left(\mu(x)y\right)}{dx} = \mu(x)\frac{dy}{dx} + \frac{d\mu(x)}{dx}y$$

3. Then identify the right side of the equation in Step 2 with the left side of the equation in Step 1:

$$\frac{d\mu(x)}{dx} = 2\mu(x)$$

4. Rearrange the terms to get

$$\frac{d\mu(x)}{\mu(x)} = 2dx$$

5. Then integrate:

$$\ln |\mu(x)| = 2x + b$$

6. Raise $e$ to the power of both sides (where $c = e^b$) to get

$$\mu(x) = ce^{2x}$$

7. Multiply the original differential equation by the integrating factor (canceling out $c$):

$$e^{2x}\frac{dy}{dx} + 2e^{2x}y = 14e^{2x}$$

8. Then combine the terms on the left side of this equation to get

$$\frac{d(e^{2x}y)}{dx} = 14e^{2x}$$

9. Next, multiply by $dx$:

$$d(e^{2x}y) = 14e^{2x}\,dx$$

10. Integrate:

$$e^{2x}y = 7e^{2x} + c$$

11. Then divide both sides by $e^{2x}$ to get

$$y = 7 + ce^{-2x}$$

12. Finally, apply the initial condition to achieve your answer:

$$y = 7 + 2e^{-2x}$$

**16**  **In the following differential equation, find $y$ by using an integrating factor:**

$$\frac{dy}{dx} + 9y = 63$$

**where**

$$y(0) = 8$$

**Solution: $y = 7 + e^{-9x}$**

1. Multiply both sides by $\mu(x)$:

$$\mu(x)\frac{dy}{dx} + 9\mu(x)y = 63\mu(x)$$

2. Identify the left side of the equation with a derivative (in this case, the derivative of a product):

$$\frac{d(\mu(x)y)}{dx} = \mu(x)\frac{dy}{dx} + \frac{d\mu(x)}{dx}y$$

3. Identify the right side of the equation in Step 2 with the left side of the equation in Step 1:

$$\frac{d\mu(x)}{dx} = 9\mu(x)$$

4. Next, rearrange the terms:

$$\frac{d\mu(x)}{\mu(x)} = 9dx$$

5. Integrate to get

$$\ln |\mu(x)| = 9x + b$$

6. Raise $e$ to the power of both sides (where $c = e^b$)

$$\mu(x) = ce^{9x}$$

7. Multiply the original equation by the integrating factor (canceling out $c$) to get

$$e^{9x}\frac{dy}{dx} + 9e^{9x}y = 63e^{9x}$$

8. Combine terms on the left side of the equation:

$$\frac{d(e^{9x}y)}{dx} = 63e^{9x}$$

9. Multiply by $dx$:

$$d(e^{9x}y) = 63e^{9x}\,dx$$

10. Then integrate to get

$$e^{9x}y = 7e^{9x} + c$$

11. Next, divide both sides by $e^{9x}$:

$$y = 7 + ce^{-9x}$$

12. After applying the initial condition, you should have

$$y = 7 + e^{-9x}$$

# Chapter 2

# Surveying Separable First Order Differential Equations

## In This Chapter

▶ Diving into separable differential equations

▶ Knowing how to obtain implicit solutions

▶ Practicing the $y = vx$ trick for separating differential equations

▶ Solving separable first order differential equations with initial conditions

*W*elcome to separable differential equations! You know 'em; you may even love 'em. After all, they let you separate out the variables so only one variable appears on each side of the equal sign. What's not to love about that?

In this chapter, you're not going to limit yourself to linear differential equations (like those covered in Chapter 1). That is, you may see something like this:

$$\frac{dy}{dx} + ay^2 = g(x)$$

But because the equations in this chapter are still considered first order, you can expect to see something along these lines:

$$M(x, y) + N(x, y)\,\frac{dy}{dx} = 0$$

To restrict the form of this differential equation even more, say that $M(x, y)$ is really just a function of $x$ — that is, $M(x)$. Similarly, say that $N(x, y)$ is really just a function of $y$ — that is, $N(y)$. Combined, that gives you

$$M(x) + N(y)\,\frac{dy}{dx} = 0$$

This differential equation is considered separable, because it can be written in a form where all terms in $x$ are on one side of the equal sign and all terms in $y$ are on the other side.

For instance, multiplying by $dx$ gives you

$$M(x)\,dx + N(y)\,dy = 0$$

and that can be written as

$$M(x)\,dx = -N(y)\,dy$$

which means you've separated the differential equation so that only $x$ appears on one side and only $y$ appears on the other.

Now that that's settled, check out the following sections. They help you find implicit solutions, make the supposedly inseparable separable, and get a handle on how initial conditions affect a separable differential equation.

# The Ins and Outs of Working with Separable Differential Equations

If you can separate a differential equation, you're that much closer to solving it. Here's the general form of a separable first order differential equation:

$$M(x) + N(y)\,\frac{dy}{dx} = 0$$

Note that both $M(x)$ and $N(y)$ don't have to be linear in $x$ and $y$, respectively. For example, you may encounter this differential equation, which is separable but not linear:

$$x + y^2\,\frac{dy}{dx} = 0$$

You can separate this equation like so:

$$x\,dx + y^2\,dy = 0$$

which gives you

$$x\,dx = -y^2\,dy$$

As you can see, the resulting equation is clearly separated.

Now consider this differential equation:

$$\frac{dy}{dx} - x^2 = 0$$

where

$$y(0) = 0$$

You can separate this one into

$$dy = x^2\, dx$$

Integrating both sides gives you the following:

$$y = \frac{x^3}{3} + c$$

When you apply the initial condition, $y(0) = 0$, you get

$$c = 0$$

So the answer is

$$y = \frac{x^3}{3}$$

Here's another example of a typical separable first order differential equation, followed by some practice problems you can work out for yourself.

**Q.** Solve this differential equation:

$$\frac{dy}{dx} - x - x^2 = 0$$

where

$$y(0) = 1$$

**A.** $y = \dfrac{x^2}{2} + \dfrac{x^3}{3} + 1$

1. Multiply both sides by $dx$:

$$dy - x\, dx - x^2\, dx = 0$$

2. Separate $x$ and $y$ terms on different sides of the equal sign:

$$dy = x\, dx + x^2\, dx$$

3. Then integrate to get

$$y = \frac{x^2}{2} + \frac{x^3}{3} + c$$

4. Apply the initial conditions to find that

$$c = 1$$

5. Tada! The solution is

$$y = \frac{x^2}{2} + \frac{x^3}{3} + 1$$

**1.** Solve this differential equation:

$$\frac{dy}{dx} - x^3 = 0$$

where

$$y(0) = 5$$

*Solve It*

**2.** Figure out the answer to the following:

$$\frac{dy}{dx} - \cos(x) = 0$$

where

$$y(0) = 1$$

*Solve It*

**3.** What's the solution to this equation?

$$y\frac{dy}{dx} - x = 0$$

where

$$y(0) = 0$$

*Solve It*

**4.** Solve this differential equation:

$$\frac{dy}{dx} - x^3 - x^4 = 0$$

where

$$y(0) = 0$$

*Solve It*

**5.** Figure out the answer to the following:

$$\cos\left(y\right)\frac{dy}{dx} - x = 0$$

where

$$y(0) = 0$$

*Solve It*

**6.** What's the solution to this equation?

$$e^{y}\frac{dy}{dx} - 2x = 0$$

where

$$y(0) = \ln(2)$$

*Solve It*

# Finding Implicit Solutions

Separating differential equations into $x$ and $y$ parts is fine; it can also be quite helpful. Yet sometimes you just can't come up with a neat $y = f(x)$ solution, no matter how hard you try. For example, what if you encounter a differential equation like this one?

$$\frac{dy}{dx} = \frac{x^2}{1-y^2}$$

You can multiply both sides by $dx$ to get

$$(1 - y^2)\,dy = x^2\,dx$$

As you can see, this is a separable differential equation. Integrating both sides gives you

$$y - \frac{y^3}{3} = \frac{x^3}{3} + c$$

Doesn't exactly look easy to write in the form $y = f(x)$, does it? That's because it's an *implicit solution,* also known as any solution you can't write like $y = f(x)$. (Solutions that can be written the easy way are considered *explicit solutions.*)

Although finding implicit solutions can be useful, sometimes you end up having to resort to numerical methods on a computer to convert them into the standard $y = f(x)$ form. On the other hand, finding an implicit solution is occasionally the very best you can do.

Check out how to solve the following implicit solution problem and then try your hand at a few that are just like it.

**Q.** Find an implict solution to this differential equation:

$$\left(y - y^2\right) \frac{dy}{dx} - x^2 = 0$$

**A.** $\dfrac{y^2}{2} - \dfrac{y^3}{3} = \dfrac{x^3}{3} + c$

1. Multiply both sides by $dx$:

$$(y - y^2)\, dy - x^2\, dx = 0$$

2. Separate $x$ and $y$ terms on different sides of the equal sign:

$$(y - y^2)\, dy = x^2\, dx$$

3. Integrate to get

$$\frac{y^2}{2} - \frac{y^3}{3} = \frac{x^3}{3} + c$$

**7.** Find the implicit solution to this differential equation:

$$\left(y^2 + y\right) \frac{dy}{dx} - x = 0$$

*Solve It*

**8.** What's the implicit solution to this equation?

$$\left(y^3 + y^2 + y\right) \frac{dy}{dx} - x^2 - x = 0$$

*Solve It*

**9.** Find the implicit solution to this differential equation:

$$\sin\left(y\right)\frac{dy}{dx}-\cos\left(x\right)=0$$

Solve It

**10.** What's the implicit solution to this equation?

$$\left(e^{y}+3y^{2}\right)\frac{dy}{dx}-2x=0$$

Solve It

# Getting Tricky: Separating the Seemingly Inseparable

Sometimes you can convert differential equations that don't look separable into separable ones by using a cool trick. Why would you want to take the time? Because separable equations are usually much easier to solve than differential equations that don't appear separable.

To work some conversion magic on a differential equation, simply substitute $y = vx$ into the equation. Often the result is an easier-to-solve separable equation.

Using $y = vx$ is a useful trick when your differential equation is of the following form:

$$\frac{dy}{dx}=f\left(x,\ y\right)$$

Note that this trick only has a hope of working if $f(x, y) = f(tx, ty)$ where $t$ is a constant (meaning when you put in $tx$ for $x$ and $ty$ for $y$, the $t$ drops out).

Take a look at this problem:

$$\frac{dy}{dx}=\frac{2y^{3}+x^{3}}{xy^{2}}$$

This differential equation may seem hopelessly inseparable to the uneducated, but lucky for you, you're armed with the $y = vx$ substitution trick!

First things first though: Make sure that $f(x, y) = f(tx, ty)$. Substituting $tx$ for $x$ and $ty$ for $y$ gives you

$$\frac{dy}{dx} = \frac{2t^3y^3 + t^3x^3}{t^3xy^2}$$

The $t$ drops out, leaving you with

$$\frac{dy}{dx} = \frac{2y^3 + x^3}{xy^2}$$

So $f(x, y) = f(tx, ty)$, which means you can try applying the $y = vx$ trick. Substituting $y = vx$ into this differential equation gives you

$$v + x\,\frac{dv}{dx} = \frac{2(xv)^3 + x^3}{x(xv)^2}$$

which becomes

$$x\,\frac{dv}{dx} = \frac{v^3 + 1}{v^2}$$

Looks like this equation can be separated. When you do that, you get

$$\frac{dx}{x} = \frac{v^2 dv}{v^3 + 1}$$

Not too shabby. Now you can integrate to get

$$\ln\left(x\right) = \frac{\ln\left(v^3 + 1\right)}{3} + k$$

where $k$ is a constant of integration. If you use the fact that

$$\ln(a) + \ln(b) = \ln(ab)$$

and

$$a\ln(b) = \ln(b^a)$$

you get

$$v^3 + 1 = (mx)^3$$

where $m$ is a constant.

This equation is still in terms of $v$ and $x$, but you want a solution in terms of $x$ and $y$. Start substituting! From

$$v = {}^y\!/_x$$

you get

$$(y/x)^3 + 1 = (mx)^3$$

or

$$y^3 + x^3 = m^3 x^6$$

Solving for $y$ gives you

$$y = (cx^6 - x^3)^{1/3}$$

where $c$ is a constant. And that's the solution. Pretty cool, huh?

Review the following problem if you want to see another example of how to apply this handy trick that will make you the envy of all your friends (well, probably not really, but it'll definitely make your differential equations experience a little easier!). Think you have a handle on it already? Move ahead to the practice problems.

**Q.** Solve this differential equation by converting it into a separable form:

$$\frac{dy}{dx} = \frac{x^4 + 2y^4}{xy^3}$$

**A.** $y = (cx^8 - x^4)^{1/4}$

1. First, test whether $f(x, y) = f(tx, ty)$. Substituting $tx$ for $x$ and $ty$ for $y$ gives you

$$\frac{dy}{dx} = \frac{t^4 x^4 + 2t^4 y^4}{tx\, t^3 y^3}$$

Because the $t$ drops out, you can try the $y = vx$ trick.

2. Substituting $y = vx$ into the equation gives you

$$v + x\frac{dv}{dx} = \frac{x^4 + 2v^4 x^4}{x(xv)^3}$$

which becomes

$$x\frac{dv}{dx} = \frac{v^4 + 1}{v^3}$$

3. Separate this result and get

$$\frac{dx}{x} = \frac{v^3 dv}{v^4 + 1}$$

4. Then integrate (where $k$ is a constant of integration):

$$\ln\left(x\right) = \frac{\ln\left(v^4 + 1\right)}{4} + k$$

5. Using the fact that

$$\ln\left(a\right) + \ln\left(b\right) = \ln\left(ab\right)$$

and

$$a \ln\left(b\right) = \ln\left(b^a\right)$$

you get

$$v^4 + 1 = (mx)^4$$

where $m$ is a constant.

6. Next, substitute $v = y/x$:

$$(y/x)^4 + 1 = (mx)^4$$

or

$$y^4 + x^4 = m^4 x^8$$

7. Last, but certainly not least, solve for $y$ to get

$$y = (mx^8 - x^4)^{1/4}$$

where $m$ is a constant.

**11.** Solve this differential equation by separating it:

$$\frac{dy}{dx} = \frac{2x^4 + 2y^4}{xy^3}$$

Solve It

**12.** Convert this equation so that it's separable and solve:

$$\frac{dy}{dx} = \frac{x^2 + 2y^2}{xy}$$

Solve It

**13.** Solve this differential equation by separating it:

$$\frac{dy}{dx} = \frac{x^3 + 2y^3}{xy^2}$$

Solve It

**14.** Convert this equation so that it's separable and solve:

$$\frac{dy}{dx} = \frac{5x^5 + 2y^5}{xy^4}$$

Solve It

# Practicing Your Separation Skills

Here's your chance to practice some general separable first order differential equations. (I promise it'll be more fun than stabbing yourself in the eye, so put that pencil back down.) I include differential equations of all sorts so you can get some practice with the various possibilities. But first, a quick example.

**Q.** Solve this differential equation:

$$\frac{dy}{dx} = \frac{x^2}{y}$$

**A.** $y = \left(2\frac{x^3}{3} + c\right)^{1/2}$

1. Separate the equation to get

$$y\,\frac{dy}{dx} = x^2$$

2. Multiply both sides by $dx$:

$$y\,dy = x^2\,dx$$

3. Then integrate both sides of the equation:

$$\frac{y^2}{2} = \frac{x^3}{3} + c$$

4. Multiply by 2 (and absorb 2 into the constant $c$):

$$y^2 = 2\frac{x^3}{3} + c$$

5. You're left with

$$y = \left(2\frac{x^3}{3} + c\right)^{1/2}$$

---

**15.** Solve this differential equation by separating it:

$$\frac{dy}{dx} = \frac{x^2}{y\left(1 + x^3\right)}$$

*Solve It*

**16.** What's the solution if you separate this equation?

$$\frac{dy}{dx} - y^2 \sin\left(x\right) = 0$$

*Solve It*

**17.** Find the answer to this equation by separating it:

$$\frac{dy}{dx} = \frac{\left(4x^3 - 1\right)}{y}$$

Solve It

**18.** Solve this differential equation by separating it:

$$x\,\frac{dy}{dx} = \left(1 - y^2\right)^{1/2}$$

Solve It

**19.** What's the solution if you separate this equation?

$$\frac{dy}{dx} = \frac{\left(5x^4 - 1\right)}{y^2}$$

Solve It

**20.** Find the answer to this equation by separating it:

$$\frac{dy}{dx} = \frac{x^2}{1 + y^4}$$

Solve It

# An Initial Peek at Separable Equations with Initial Conditions

It's a given in the world of differential equations: You're going to run into separable first order differential equations with initial conditions. Having to solve a problem with an initial condition adds another dimension to the problem, as you can see in the following example and practice problems.

**Q.** Solve this differential equation:

$$\frac{dy}{dx} = \frac{x^4}{y^2}$$

where

$$y(0) = 2$$

**A.** $y = \left(3\dfrac{x^5}{5} + 8\right)^{1/3}$

1. Multiply both sides of the equation by $y^2$ to get

$$y^2 \frac{dy}{dx} = x^4$$

2. Then multiply both sides by $dx$:

$$y^2 \, dy = x^4 \, dx$$

3. Integrating both sides gives you

$$\frac{y^3}{3} = \frac{x^5}{5} + c$$

4. Multiply by 3 (and absorb 3 into the constant $c$):

$$y^3 = 3\frac{x^5}{5} + c$$

5. Then take the cube root:

$$y = \left(3\frac{x^5}{5} + c\right)^{1/3}$$

6. Solve for $c$:

$$2 = (c)^{1/3}$$

7. Here's your solution:

$$y = \left(3\frac{x^5}{5} + 8\right)^{1/3}$$

**21.** Figure out the answer to this differential equation:

$$\frac{dy}{dx} = \frac{x^5}{y^3}$$

where

$$y(0) = 3$$

*Solve It*

**22.** Solve this equation:

$$\frac{dy}{dx} = \frac{4x^3 + 6x^2}{y^2}$$

where

$$y(1) = 3$$

*Solve It*

**23.** Figure out the answer to this differential equation:

$$\frac{dy}{dx} = (1 - 2x)y^2$$

where

$$y(0) = 1$$

*Solve It*

**24.** Solve this equation:

$$1 - e^{-x}\frac{dy}{dx} = 0$$

where

$$y(0) = 2$$

*Solve It*

# Answers to Separable First Order Differential Equation Problems

Here are the answers to the practice questions I provide throughout this chapter. I walk you through each answer so you can see the problems worked out step by step. Enjoy!

**1** **Solve this differential equation:**

$$\frac{dy}{dx} - x^3 = 0$$

**where**

$$y(0) = 5$$

**Solution:** $y = \dfrac{x^4}{4} + 5$

1. First, multiply both sides by $dx$:

   $dy - x^3\, dx = 0$

2. Separate $x$ and $y$ terms on different sides of the equal sign:

   $dy = x^3\, dx$

3. Then integrate to get

   $y = \dfrac{x^4}{4} + c$

4. Last but not least, apply the initial condition to find that

   $c = 5$

5. So your answer is

   $y = \dfrac{x^4}{4} + 5$

**2** **Figure out the answer to the following:**

$$\frac{dy}{dx} - \cos\left(x\right) = 0$$

**where**

$$y(0) = 1$$

**Solution:** $y = \sin(x) + 1$

1. Multiply both sides by $dx$:

   $dy - \cos(x)\, dx = 0$

2. Then separate $x$ and $y$ terms on opposite sides of the equal sign:

   $dy = \cos(x)\, dx$

3. Next up, integrate:

   $y = \sin(x) + c$

4. Applying the initial condition tells you that

$c = 1$

5. So the solution is

$y = \sin(x) + 1$

**3** **What's the solution to this equation?**

$$y\,\frac{dy}{dx} - x = 0$$

where

$y(0) = 0$

Solution: $y = x$

1. First, multiply both sides by $dx$:

$y\,dy - x\,dx = 0$

2. Separate $x$ and $y$ terms on different sides of the equal sign:

$y\,dy = x\,dx$

3. Then integrate to get

$$\frac{y^2}{2} = \frac{x^2}{2} + c$$

4. Apply the initial condition to find that

$c = 0$

which means

$$\frac{y^2}{2} = \frac{x^2}{2}$$

5. Multiply by 2 to get

$y^2 = x^2$

6. So your answer is

$y = x$

**4** **Solve this differential equation:**

$$\frac{dy}{dx} - x^3 - x^4 = 0$$

where

$y(0) = 0$

Solution: $y = \dfrac{x^4}{4} + \dfrac{x^5}{5}$

1. Multiply both sides by $dx$:

$dy - x^3\,dx - x^4\,dx = 0$

2. Then separate $x$ and $y$ terms on opposite sides of the equal sign:

$dy = x^3\,dx + x^4\,dx$

3. Next up, integrate:

$$y = \frac{x^4}{4} + \frac{x^5}{5} + c$$

4. Applying the initial condition tells you that

$$c = 0$$

5. So the solution is

$$y = \frac{x^4}{4} + \frac{x^5}{5}$$

**5**  **Figure out the answer to the following:**

$$\cos\left(y\right) \frac{dy}{dx} - x = 0$$

**where**

$$y(0) = 0$$

**Solution:** $y = \sin^{-1}\left(\frac{x^2}{2} + n\pi\right)$  $n = 0, 1, 2...$

1. First, multiply both sides by $dx$:

$$\cos(y)\, dy - x\, dx = 0$$

2. Separate $x$ and $y$ terms on different sides of the equal sign:

$$\cos(y)\, dy = x\, dx$$

3. Then integrate to get

$$\sin\left(y\right) = \frac{x^2}{2} + c$$

4. Go ahead and take the inverse sine:

$$y = \sin^{-1}\left(\frac{x^2}{2} + c\right)$$

5. Last but not least, apply the initial condition to find that

$$c = n\pi \quad n = 0, 1, 2 \ldots$$

6. So your answer is

$$y = \sin^{-1}\left(\frac{x^2}{2} + n\pi\right) \quad n = 0, 1, 2 \ldots$$

**6**  **What's the solution to this equation?**

$$e^y \frac{dy}{dx} - 2x = 0$$

**where**

$$y(0) = \ln(2)$$

**Solution:** $y = \ln |x^2 + 2|$

1. Multiply both sides by $dx$:

   $e^y \, dy - 2x \, dx = 0$

2. Then separate $x$ and $y$ terms on opposite sides of the equal sign:

   $e^y \, dy = 2x \, dx$

3. Next up, integrate:

   $e^y = x^2 + c$

   and take the natural log:

   $y = \ln |x^2 + c|$

4. Applying the initial condition tells you that

   $c = 2$

5. So the solution is

   $y = \ln |x^2 + 2|$

**7** **Find the implicit solution to this differential equation:**

$$\left(y^2 + y\right) \frac{dy}{dx} - x = 0$$

**Solution:** $\dfrac{y^3}{3} + \dfrac{y^2}{2} = \dfrac{x^2}{2} + c$

1. Multiply both sides by $dx$:

   $(y^2 + y) \, dy - x \, dx = 0$

2. Then separate $x$ and $y$ terms on different sides of the equal sign:

   $(y^2 + y) \, dy = x \, dx$

3. Finally, integrate to get

   $$\frac{y^3}{3} + \frac{y^2}{2} = \frac{x^2}{2} + c$$

**8** **What's the implicit solution to this equation?**

$$\left(y^3 + y^2 + y\right) \frac{dy}{dx} - x^2 - x = 0$$

**Solution:** $\dfrac{y^4}{4} + \dfrac{y^3}{3} + \dfrac{y^2}{2} = \dfrac{x^3}{3} + \dfrac{x^2}{2} + c$

1. Start off by multiplying both sides of the equation by $dx$:

   $(y^3 + y^2 + y) \, dy - x^2 \, dx - x \, dx = 0$

2. Next, separate $x$ and $y$ terms on opposite sides of the equal sign:

   $(y^3 + y^2 + y) \, dy = x^2 \, dx + x \, dx$

3. Then integrate:

   $$\frac{y^4}{4} + \frac{y^3}{3} + \frac{y^2}{2} = \frac{x^3}{3} + \frac{x^2}{2} + c$$

**9**  **Find the implicit solution to this differential equation:**

$$\sin\,(y)\,\frac{dy}{dx} - \cos\,(x) = 0$$

**Solution:** $y = \cos^{-1}(\sin(x)) + C$

1. Multiply both sides by $dx$:

   $\sin(y)\,dy - \cos(x)\,dx = 0$

2. Then separate $x$ and $y$ terms on different sides of the equal sign:

   $\sin(y)\,dy = \cos(x)\,dx$

3. Integrate to get the following (where $C$ is a constant of integration):

   $\cos(y) = -\sin(x) + C$

4. The result is an explicit solution:

   $y = \cos^{-1}(\sin(x)) + C$

**10**  **What's the implicit solution to this equation?**

$$\left(e^{y} + 3y^{2}\right)\frac{dy}{dx} - 2x = 0$$

**Solution:** $e^{y} + y^{3} = x^{2}$

1. Start off by multiplying both sides of the equation by $dx$:

   $(e^{y} + 3y^{2})\,dy - 2x\,dx = 0$

2. Next, separate $x$ and $y$ terms on opposite sides of the equal sign:

   $(e^{y} + 3y^{2})\,dy = 2x\,dx$

3. Then integrate:

   $e^{y} + y^{3} = x^{2}$

**11**  **Solve this differential equation by separating it:**

$$\frac{dy}{dx} = \frac{2x^{4} + 2y^{4}}{xy^{3}}$$

**Solution:** $y = (mx^{8} - 2x^{4})^{1/4}$

1. First, test whether $f(x, y) = f(tx, ty)$. Substituting $tx$ for $x$ and $ty$ for $y$ gives you

   $$\frac{dy}{dx} = \frac{2t^{4}x^{4} + 2t^{4}y^{4}}{tx\,t^{3}y^{3}}$$

   The $t$ drops out, which means you can try the $y = vx$ trick.

2. Substitute $y = vx$ to get

   $$v + x\,\frac{dv}{dx} = \frac{2x^{4} + 2v^{4}x^{4}}{x(xv)^{3}}$$

   which becomes

   $$x\,\frac{dv}{dx} = \frac{v^{4} + 2}{v^{3}}$$

3. Separate the terms as follows:

$$\frac{dx}{x} = \frac{v^3 dv}{v^4 + 2}$$

4. Then integrate:

$$\ln (x) = \frac{\ln (v^4 + 2)}{4} + k$$

*Note:* Here, $k$ is a constant of integration.

5. Using the fact that

$$\ln (a) + \ln (b) = \ln (ab)$$

and

$$a \ln (b) = \ln (b^a)$$

you get

$$v^4 + 2 = (mx)^4$$

where $m$ is a constant.

6. Substitute $v = {}^y/_x$:

$$({}^y/_x)^4 + 2 = (mx)^4$$

or

$$y^4 + 2x^4 = m^4 x^8$$

7. Finally, solve for $y$:

$$y = (mx^8 - 2x^4)^{1/4}$$

where $m$ is a constant.

**12** **Convert this equation so that it's separable and solve:**

$$\frac{dy}{dx} = \frac{x^2 + 2y^2}{xy}$$

**Solution:** $y = (mx^4 - x^2)^{1/2}$

1. Test whether $f(x, y) = f(tx, ty)$. Substituting $tx$ for $x$ and $ty$ for $y$ gives you

$$\frac{dy}{dx} = \frac{t^2 x^2 + 2t^2 y^2}{tx \, ty}$$

Because the $t$ drops out, you can employ the $y = vx$ trick.

2. Substituting $y = vx$ gives you

$$v + x \frac{dv}{dx} = \frac{x^2 + 2v^2 x^2}{x(xv)}$$

That equation becomes

$$x \frac{dv}{dx} = \frac{v^2 + 1}{v}$$

3. Separate the $x$ and $v$ variables:

$$\frac{dx}{x} = \frac{v\,dv}{v^2 + 1}$$

4. Now integrate to get

$$\ln\left(x\right) = \frac{\ln\left(v^2 + 1\right)}{2} + k$$

where $k$ is a constant of integration.

5. Using the fact that

$$\ln\left(a\right) + \ln\left(b\right) = \ln\left(ab\right)$$

and

$$a\ln\left(b\right) = \ln\left(b^a\right)$$

you wind up with

$$v^2 + 1 = (mx)^2$$

where $m$ is a constant.

6. Substituting $v = {}^y/_x$ gives you

$$({}^y/_x)^2 + 1 = (mx)^2$$

or

$$y^2 + x^2 = m^2 x^4$$

7. Now just solve for $y$ to get

$$y = (mx^4 - x^2)^{1/2}$$

where $m$ is a constant.

**13** **Solve this differential equation by separating it:**

$$\frac{dy}{dx} = \frac{x^3 + 2y^3}{xy^2}$$

**Solution:** $y = (mx^6 - x^3)^{1/3}$

1. First, test whether $f(x, y) = f(tx, ty)$. Substituting $tx$ for $x$ and $ty$ for $y$ gives you

$$\frac{dy}{dx} = \frac{t^3 x^3 + 2t^3 y^3}{tx\,t^2 y^2}$$

The $t$ drops out, which means you can try the $y = vx$ trick.

2. Substitute $y = vx$ to get

$$v + x\,\frac{dv}{dx} = \frac{1 + 2v^3}{v^2}$$

which becomes

$$x\,\frac{dv}{dx} = \frac{v^3 + 1}{v^2}$$

3. Separate the terms as follows:

$$\frac{dx}{x} = \frac{v^2 dv}{v^3 + 1}$$

4. Then integrate:

$$\ln(x) = \frac{\ln(v^3 + 1)}{3} + k$$

*Note:* Here, $k$ is a constant of integration.

5. Using the fact that

$$\ln(a) + \ln(b) = \ln(ab)$$

and

$$a \ln(b) = \ln(b^a)$$

you get

$$v^3 + 1 = (mx)^3$$

where $m$ is a constant.

6. Substitute $v = y/x$:

$$(y/x)^3 + 1 = (mx)^3$$

or

$$y^3 + x^3 = m^3 x^6$$

7. Finally, solve for $y$:

$$y = (mx^6 - x^3)^{1/3}$$

where $m$ is a constant.

**14** **Convert this equation so that it's separable and solve:**

$$\frac{dy}{dx} = \frac{5x^5 + 2y^5}{xy^4}$$

**Solution: $y = (mx^{10} - 5x^5)^{1/5}$**

1. Test whether $f(x, y) = f(tx, ty)$. Substituting $tx$ for $x$ and $ty$ for $y$ gives you

$$\frac{dy}{dx} = \frac{5t^5 x^5 + 2t^5 y^5}{tx \, t^4 y^4}$$

Because the $t$ drops out, you can employ the $y = vx$ trick.

2. Substituting $y = vx$ gives you

$$v + x \frac{dv}{dx} = \frac{5x^5 + 2v^5 x^5}{x(xv)^4}$$

That equation becomes

$$x \frac{dv}{dx} = \frac{5 + v^5}{v^4}$$

3. Separate the $x$ and $v$ variables:

$$\frac{dx}{x} = \frac{v^4 dv}{v^5 + 5}$$

4. Now integrate to get

$$\ln(x) = \frac{\ln(v^5 + 5)}{5} + k$$

where $k$ is a constant of integration.

5. Using the fact that

$$\ln(a) + \ln(b) = \ln(ab)$$

and

$$a \ln(b) = \ln(b^a)$$

you wind up with

$$v^5 + 5 = (mx)^5$$

where $m$ is a constant.

6. Substituting $v = {}^y/_x$ gives you

$$({}^y/_x)^5 + 5 = (mx)^5$$

or

$$y^5 + 5x^5 = m^5 x^{10}$$

7. Now just solve for $y$ to get

$$y = (mx^{10} - 5x^5)^{1/5}$$

where $m$ is a constant.

**15** **Solve this differential equation by separating it:**

$$\frac{dy}{dx} = \frac{x^2}{y(1 + x^3)}$$

**Solution:** $y = \left( 2\dfrac{\ln|1 + x^3|}{3} + c \right)^{1/2}$

1. Multiply both sides by $y$:

$$y\frac{dy}{dx} = \frac{x^2}{(1 + x^3)}$$

2. Then multiply both sides by $dx$:

$$y\, dy = \frac{x^2\, dx}{(1 + x^3)}$$

3. Integrate to get

$$\frac{y^2}{2} = \frac{\ln|1 + x^3|}{3} + c$$

4. Multiply by 2:

$$y^2 = 2\frac{\ln\left|1+x^3\right|}{3} + c$$

5. Finally, take the square root to get

$$y = \left(2\frac{\ln\left|1+x^3\right|}{3} + c\right)^{1/2}$$

**16** **What's the solution if you separate this equation?**

$$\frac{dy}{dx} - y^2 \sin\left(x\right) = 0$$

**Solution:** $y = (\cos\left(x\right) + c)^{-1}$

1. Divide both sides of the equation by $y^2$:

$$\frac{1}{y^2}\frac{dy}{dx} - \sin\left(x\right) = 0$$

2. Next, multiply both sides by $dx$:

$$\frac{dy}{y^2} - \sin\left(x\right) dx = 0$$

3. Separate the two terms to get

$$\frac{dy}{y^2} = \sin\left(x\right) dx$$

4. Then integrate:

$$y^{-1} = \cos\left(x\right) + c$$

5. Finally, take the reciprocal:

$$y = (\cos\left(x\right) + c)^{-1}$$

**17** **Find the answer to this equation by separating it:**

$$\frac{dy}{dx} = \frac{\left(4x^3 - 1\right)}{y}$$

**Solution:** $y = (2x^4 - 2x + c)^{1/2}$

1. Multiply both sides by $y$:

$$y\frac{dy}{dx} = \left(4x^3 - 1\right)$$

2. Then multiply both sides by $dx$:

$$y\, dy = (4x^3 - 1)\, dx$$

3. Integrate to get

$$\frac{y^2}{2} = x^4 - x + c$$

4. Multiply by 2 (and absorb 2 into the constant $c$):

$$y^2 = 2x^4 - 2x + c$$

5. Finally, take the square root to get

$$y = (2x^4 - 2x + c)^{1/2}$$

**18** **Solve this differential equation by separating it:**

$$x \frac{dy}{dx} = \left(1 - y^2\right)^{1/2}$$

**Solution: $y = \sin\left(\ln |x| + c\right)$**

1. Divide both sides of the equation by $x$:

$$\frac{dy}{dx} = \frac{\left(1 - y^2\right)^{1/2}}{x}$$

2. Next, multiply both sides by $dx$:

$$dy = \frac{\left(1 - y^2\right)^{1/2}}{x} dx$$

3. Divide both sides by $(1 - y^2)^{1/2}$:

$$\frac{dy}{\left(1 - y^2\right)^{1/2}} = \frac{dx}{x}$$

4. Then integrate:

$$\sin^{-1}(y) = \ln |x| + c$$

5. Finally, take the sine of both sides:

$$y = \sin\left(\ln |x| + c\right)$$

**19** **What's the solution if you separate this equation?**

$$\frac{dy}{dx} = \frac{\left(5x^4 - 1\right)}{y^2}$$

**Solution: $y = (3x^5 - 3x + c)^{1/3}$**

1. Multiply both sides by $y^2$:

$$y^2 \frac{dy}{dx} = \left(5x^4 - 1\right)$$

2. Then multiply both sides by $dx$:

$$y^2 \, dy = (5x^4 - 1) \, dx$$

3. Integrate to get

$$\frac{y^3}{3} = x^5 - x + c$$

4. Multiply both sides by 3 (and absorb 3 into the constant $c$):

$y^3 = 3x^5 - 3x + c$

5. Finally, take the cube root to get

$y = (3x^5 - 3x + c)^{1/3}$

**20** **Find the answer to this equation by separating it:**

$$\frac{dy}{dx} = \frac{x^2}{1+y^4}$$

**Solution:** $y + \dfrac{y^5}{5} = \dfrac{x^3}{3} + c$

1. Multiply both sides of the equation by $1 + y^4$:

$$\left(1 + y^4\right) \frac{dy}{dx} = x^2$$

2. Next, multiply both sides by $dx$:

$(1 + y^4)\, dy = x^2\, dx$

3. Then integrate:

$$y + \frac{y^5}{5} = \frac{x^3}{3} + c$$

Leave this one as an implicit solution.

**21** **Figure out the answer to this differential equation:**

$$\frac{dy}{dx} = \frac{x^5}{y^3}$$

where

$y(0) = 3$

**Solution:** $y = \left(2\dfrac{x^6}{3} + 81\right)^{1/4}$

1. Start by multiplying both sides of the equation by $y^3$ to get

$$y^3 \frac{dy}{dx} = x^5$$

2. Multiplying both sides by $dx$ gives you

$y^3\, dy = x^5\, dx$

3. Go ahead and integrate both sides of the equation:

$$\frac{y^4}{4} = \frac{x^6}{6} + c$$

4. Then multiply by 4 (and absorb 4 into the constant $c$):

$$y^4 = 2\frac{x^6}{3} + c$$

5. Take the fourth root:

$$y = \left(2\frac{x^6}{3} + c\right)^{1/4}$$

6. Solve for $c$:

$$3 = (c)^{1/4}$$

7. Here's your solution:

$$y = \left(2\frac{x^6}{3} + 81\right)^{1/4}$$

**22** **Solve this equation:**

$$\frac{dy}{dx} = \frac{4x^3 + 6x^2}{y^2}$$

**where**

$$y(1) = 3$$

**Solution:** $y = (3x^4 + 6x^3 + 18)^{1/3}$

1. Multiply both sides by $y^2$ to get

$$y^2 \frac{dy}{dx} = 4x^3 + 6x^2$$

2. Next, multiply both sides by $dx$:

$$y^2\,dy = (4x^3 + 6x^2)\,dx$$

3. Integrate both sides:

$$\frac{y^3}{3} = x^4 + 6\frac{x^3}{3} + c$$

4. Then multiply by 3 (and absorb 3 into the constant $c$):

$$y^3 = 3x^4 + 6x^3 + c$$

5. Take the cube root:

$$y = (3x^4 + 6x^3 + c)^{1/3}$$

6. Now solve for $c$ using the initial condition:

$$3 = (9 + c)^{1/3}$$

which gives you this value for $c$:

$$c = 18$$

resulting in this solution:

$$y = (3x^4 + 6x^3 + 18)^{1/3}$$

**23** **Figure out the answer to this differential equation:**

$$\frac{dy}{dx} = \left(1 - 2x\right)y^2$$

**where**

$$y(0) = 1$$

**Solution:** $y = (x^2 - x + 1)^{-1}$

1. Start by dividing both sides of the equation by $y^2$:

$$\frac{1}{y^2} \frac{dy}{dx} = (1 - 2x)$$

2. Multiplying both sides by $dx$ gives you

$$\frac{dy}{y^2} = (1 - 2x) dx$$

3. Go ahead and integrate both sides of the equation:

$$\frac{-1}{y} = x - x^2 + c$$

4. Then take the reciprocal to get

$$-y = (x - x^2 + c)^{-1}$$

5. Multiply by $-1$:

$$y = -(x - x^2 + c)^{-1}$$

6. Solve for $c$ using the initial condition:

$$1 = -(c)^{-1}$$

7. You should wind up with this value for $c$:

$$c = -1$$

8. Here's your solution:

$$y = -(x - x^2 - 1)^{-1}$$

which you can also write as

$$y = (x^2 - x + 1)^{-1}$$

**24** **Solve this equation:**

$$1 - e^{-x} \frac{dy}{dx} = 0$$

**where**

$$y(0) = 2$$

**Solution:** $y = (2e^x + 2)^{1/2}$

1. Multiply both sides by $e^x$ to get

$$e^x - y \frac{dy}{dx} = 0$$

2. Next, multiply both sides by $dx$:

$$e^x dx - y \, dy = 0$$

3. Separate the terms:

$$e^x dx = y \, dy$$

4. Then integrate both sides:

$$e^x + c = \frac{y^2}{2}$$

5. Multiply both sides by 2 (and absorb 2 into the constant $c$):

   $$2e^x + c = y^2$$

6. Take the square root:

   $$y = (2e^x + c)^{1/2}$$

7. Now solve for $c$ using the initial condition:

   $$2 = (2 + c)^{1/2}$$

   which gives you this value for $c$:

   $$c = 2$$

   resulting in this solution:

   $$y = (2e^x + 2)^{1/2}$$

# Chapter 3

# Examining Exact First Order Differential Equations

**In This Chapter**

▶ Figuring out whether a differential equation is exact

▶ Solving exact differential equations

A n *exact* differential equation is a differential equation that can be cast into the
following form:

$$\frac{df(x,\ y)}{dx} = 0$$

You can integrate the derivative in this case and get the solution $f(x, y)$, which is precisely
what this chapter is all about: finding $f(x, y)$. In the following pages, you practice a simple
test to determine whether a differential equation is exact (so that you don't waste your time
looking for $f(x, y)$ if it's not), and then you work on solving exact differential equations.

## Exactly, Dear Watson: Determining whether a Differential Equation Is Exact

Unlike a detective who has to spend time (sometimes lots of it!) analyzing evidence, you can
figure out whether a particular differential equation is exact or not in a flash. The approach
I show you in this section saves you a good deal of time in the long run because you don't
have to invest any effort in trying to find the function $f(x, y)$ if an equation isn't exact.

Here's the test for determining whether a differential equation is exact:

If you have the differential equation

$$M(x,\ y) + N(x,\ y)\ \frac{dy}{dx} = 0$$

then there exists a function $f(x, y)$ such that

$$\frac{\partial f(x, y)}{\partial x} = M(x, y)$$

and

$$\frac{\partial f(x, y)}{\partial y} = N(x, y)$$

if and only if

$$\frac{\partial M(x, y)}{\partial y} = \frac{\partial N(x, y)}{\partial x}$$

Following is an example to illustrate this test in action. I suggest you take a few minutes to review it before diving into the following practice problems that ask you to determine whether a differential equation is exact or not.

**Q.** Is this differential equation exact?

$$2x + y^2 + 2xy \frac{dy}{dx} = 0$$

**A.** Yes.

1. To solve this equation, you first need to put it into this form:

$$M(x, y) + N(x, y) \frac{dy}{dx} = 0$$

2. So

$$M(x, y) = 2x + y^2$$

and

$$N(x, y) = 2xy$$

3. Now calculate the following equations:

$$\frac{\partial M(x, y)}{\partial y} = 2y$$

$$\frac{\partial N(x, y)}{\partial x} = 2y$$

4. So

$$\frac{\partial M(x, y)}{\partial y} = \frac{\partial N(x, y)}{\partial x} = 2y$$

Therefore, the differential equation is exact.

**1.** Is this differential equation exact?

$$y^2 + 2xy \, \frac{dy}{dx} = 0$$

Solve It

**2.** Determine whether this differential equation is exact:

$$y + xy \, \frac{dy}{dx} = 0$$

Solve It

**3.** Following is an exact differential equation . . . or is it? Calculate to decide.

$$5y + 10x + 5x \, \frac{dy}{dx} = 0$$

Solve It

**4.** Is this differential equation exact?

$$-2y - 2x \, \frac{dy}{dx} = 0$$

Solve It

**5.** Determine whether this differential equation is exact:

$$\frac{1}{y^2} - \frac{x}{y^2}\frac{dy}{dx} = 0$$

*Solve It*

**6.** Following is an exact differential equation . . . or is it? Calculate to decide.

$$xy + \frac{x^2}{2}\frac{dy}{dx} = 0$$

*Solve It*

**7.** Is this differential equation exact?

$$\frac{1}{y^2} - \frac{3x}{y^3}\frac{dy}{dx} = 0$$

*Solve It*

**8.** Determine whether this differential equation is exact:

$$y^2 + 1 + xy\frac{dy}{dx} = 0$$

*Solve It*

# Getting Answers from Exact Differential Equations

To find the solution to an exact differential equation like this one:

$$M(x,\ y) + N(x,\ y)\ \frac{dy}{dx} = 0$$

in a systematic way, you need to find a function $f(x, y)$ such that

$$\frac{\partial f(x,\ y)}{\partial x} = M(x,\ y)$$

and

$$\frac{\partial f(x,\ y)}{\partial y} = N(x,\ y)$$

The correct function allows you to write the differential equation like this:

$$\frac{\partial f(x,\ y)}{\partial x} + \frac{\partial f(x,\ y)}{\partial y}\ \frac{dy}{dx} = 0$$

Here's your assignment, should you choose to accept it: Try to integrate $M(x, y)$ and $N(x, y)$ with respect to $x$ and $y$ to see whether you can find $f(x, y)$.

$$2xy + (1 + x^2)\ \frac{dy}{dx} = 0$$

In this case,

$$M(x, y) = 2xy$$

which means that

$$\frac{\partial f(x,\ y)}{\partial x} = 2xy$$

If you integrate both sides, you get

$$f(x, y) = x^2 y + g(y)$$

where $g(y)$ is a function that depends only on $y$, not on $x$. What's $g(y)$? Ah, but you already know the answer to that question because

$$\frac{\partial f(x,\ y)}{\partial y} = N(x,\ y)$$

and

$$N(x, y) = (1 + x^2)$$

so

$$\frac{\partial f(x,\ y)}{\partial y} = \left(1 + x^2\right)$$

Because you know that

$$f(x,\ y) = x^2 y + g(y)$$

you get this:

$$\frac{\partial f(x,\ y)}{\partial y} = x^2 + \frac{\partial g(y)}{\partial y} = 1 + x^2$$

So

$$\frac{\partial g(y)}{\partial y} = 1$$

Integrating with respect to $y$ gives you the following:

$$g(y) = y + k$$

where $k$ is a constant of integration. And because

$$f(x,\ y) = x^2 y + g(y)$$

you get

$$f(x,\ y) = x^2 y + y + k$$

You now know that the solution is

$$f(x,\ y) = c$$

which can be rewritten as

$$x^2 y + y = c$$

Note that $k$, the constant of integration, has been absorbed into the constant $c$. That's the implicit solution to the exact differential equation. Want the explicit solution instead? Here you go:

$$y = \frac{c}{\left(1 + x^2\right)}$$

The following example walks you through this process again, but feel free to skip ahead to the practice problems if you think you're ready to test out your skills.

**Q.** Solve this exact differential equation:

$$y + x\,\frac{dy}{dx} = 0$$

**A.** $y = c/x$

1. You can identify that

$$M(x, y) = y$$

or

$$\frac{\partial f(x, y)}{\partial x} = y$$

2. Integrate to get

$$f(x, y) = xy + g(y)$$

where $g(y)$ is a function.

3. Do the same with the second part of the equation. Identify that

$$N(x, y) = x$$

Because

$$\frac{\partial f(x, y)}{\partial y} = N(x, y)$$

that means

$$\frac{\partial f(x, y)}{\partial y} = x$$

4. If

$$\frac{\partial f(x, y)}{\partial y} = x + \frac{\partial g(y)}{\partial y}$$

then

$$\frac{\partial g(y)}{\partial y} = 0$$

5. Integrate again to get

$$g(y) = k$$

where $k$ is a constant of integration.

6. Because

$$f(x, y) = xy + g(y)$$

you get

$$f(x, y) = xy + k$$

7. In general, the solution is

$$f(x, y) = c$$

so

$$xy + k = c$$

8. Absorbing $k$ into $c$ gives you

$$y = c/x$$

**9.** What's the solution to this exact differential equation?

$$3x^2 + 2y \frac{dy}{dx} = 0$$

Solve It

**10.** Solve this exact differential equation:

$$\frac{1}{y^2} - \frac{2x}{y^3} \frac{dy}{dx} = 0$$

Solve It

**11.** What's the solution to this exact differential equation?

$$y + 2x + x \frac{dy}{dx} = 0$$

Solve It

**12.** Solve this exact differential equation:

$$\frac{1}{y} - \frac{x}{y^2} \frac{dy}{dx} = 0$$

Solve It

# Answers to Exact First Order Differential Equation Problems

Following are the answers to the practice questions presented throughout this chapter. Each one is worked out step by step so that if you messed one up along the way, you can more easily see where you took a wrong turn.

**1** **Is this differential equation exact?**

$$y^2 + 2xy \, \frac{dy}{dx} = 0$$

**Answer: Yes**

1. Start by casting the differential equation in this form:

$$M(x, \ y) + N(x, \ y) \, \frac{dy}{dx} = 0$$

2. So

$M(x, y) = y^2$

and

$N(x, y) = 2xy$

3. Now calculate the following equations:

$$\frac{\partial M(x, \ y)}{\partial y} = 2y$$

and

$$\frac{\partial N(x, \ y)}{\partial x} = 2y$$

4. So

$$\frac{\partial M(x, \ y)}{\partial y} = \frac{\partial N(x, \ y)}{\partial x} = 2y$$

Therefore, the differential equation is exact.

**2** **Determine whether this differential equation is exact:**

$$y + xy \, \frac{dy}{dx} = 0$$

**Answer: No**

1. Put the equation into the following form:

$$M(x, \ y) + N(x, \ y) \, \frac{dy}{dx} = 0$$

2. You know that

$$M(x, y) = y$$

and

$$N(x, y) = xy$$

3. Consequently, you can calculate that

$$\frac{\partial M(x,\ y)}{\partial y} = 1$$

and

$$\frac{\partial N(x,\ y)}{\partial y} = y$$

to get

$$\frac{\partial M(x,\ y)}{\partial y} \neq \frac{\partial N(x,\ y)}{\partial x}$$

Thus, the differential equation must *not* be exact.

**3**   **Following is an exact differential equation . . . or is it? Calculate to decide.**

$$5y + 10x + 5x\ \frac{dy}{dx} = 0$$

**Answer: Yes**

1. Start by casting the differential equation in this form:

$$M(x,\ y) + N(x,\ y)\ \frac{dy}{dx} = 0$$

2. So

$$M(x, y) = 5y + 10x$$

and

$$N(x, y) = 5x$$

3. Now calculate the following equations:

$$\frac{\partial M(x,\ y)}{\partial y} = 5$$

and

$$\frac{\partial N(x,\ y)}{\partial x} = 5$$

4. So

$$\frac{\partial M(x,\ y)}{\partial y} = \frac{\partial N(x,\ y)}{\partial x} = 5$$

Therefore, the differential equation is exact.

**4**  **Is this differential equation exact?**

$$-2y - 2x\,\frac{dy}{dx} = 0$$

**Answer: Yes**

1. Put the equation into the following form:

$$M(x,\ y) + N(x,\ y)\,\frac{dy}{dx} = 0$$

2. You know that

$$M(x, y) = -2y$$

and

$$N(x, y) = -2x$$

3. Consequently, you can calculate that

$$\frac{\partial M(x,\ y)}{\partial y} = -2$$

and

$$\frac{\partial N(x,\ y)}{\partial x} = -2$$

to get

$$\frac{\partial M(x,\ y)}{\partial y} = \frac{\partial N(x,\ y)}{\partial x} = -2$$

Thus, the differential equation is exact.

**5**  **Determine whether this differential equation is exact:**

$$\frac{1}{y^2} - \frac{x}{y^2}\,\frac{dy}{dx} = 0$$

**Answer: No**

1. Start by casting the differential equation in this form:

$$M(x,\ y) + N(x,\ y)\,\frac{dy}{dx} = 0$$

2. So

$$M(x,\ y) = \frac{1}{y^2}$$

and

$$N(x,\ y) = -\frac{x}{y^2}$$

3. Now calculate the following equations:

$$\frac{\partial M(x,\ y)}{\partial y} = -\frac{2}{y^3}$$

and

$$\frac{\partial N(x,\ y)}{\partial x} = -\frac{2}{y^2}$$

4. So

$$\frac{\partial M(x,\ y)}{\partial y} \neq \frac{\partial N(x,\ y)}{\partial x}$$

Therefore, the differential equation must *not* be exact.

**6** **Following is an exact differential equation . . . or is it? Calculate to decide.**

$$xy + \frac{x^2}{2}\frac{dy}{dx} = 0$$

**Answer: Yes**

1. Put the equation into the following form:

$$M(x,\ y) + N(x,\ y)\frac{dy}{dx} = 0$$

2. You know that

$$M(x, y) = xy$$

and

$$N(x,\ y) = \frac{x^2}{2}$$

3. Consequently, you can calculate that

$$\frac{\partial M(x,\ y)}{\partial y} = x$$

and

$$\frac{\partial N(x,\ y)}{\partial x} = x$$

to get

$$\frac{\partial M(x,\ y)}{\partial y} = \frac{\partial N(x,\ y)}{\partial x} = x$$

Thus, the differential equation is exact.

**7** **Is this differential equation exact?**

$$\frac{1}{y^2} - \frac{3x}{y^3} \frac{dy}{dx} = 0$$

**Answer: No**

1. Start by casting the differential equation in this form:

$$M(x,\ y) + N(x,\ y)\ \frac{dy}{dx} = 0$$

2. So

$$M(x,\ y) = \frac{1}{y^2}$$

and

$$N(x,\ y) = -\frac{3x}{y^2}$$

3. Now calculate the following equations:

$$\frac{\partial M(x,\ y)}{\partial y} = -\frac{2}{y^3}$$

and

$$\frac{\partial N(x,\ y)}{\partial x} = -\frac{3}{y^3}$$

4. So

$$\frac{\partial M(x,\ y)}{\partial y} \neq \frac{\partial N(x,\ y)}{\partial x}$$

Therefore, the differential equation must *not* be exact.

**8** **Determine whether this differential equation is exact:**

$$y^2 + 1 + xy\ \frac{dy}{dx} = 0$$

**Answer: No**

1. Put the equation into the following form:

$$M(x,\ y) + N(x,\ y)\ \frac{dy}{dx} = 0$$

2. You know that

$$M(x, y) = y^2 + 1$$

and

$$N(x, y) = xy$$

3. Consequently, you can calculate that

$$\frac{\partial M(x, y)}{\partial y} = 2y$$

and

$$\frac{\partial N(x, y)}{\partial x} = y$$

to get

$$\frac{\partial M(x, y)}{\partial y} \neq \frac{\partial N(x, y)}{\partial x}$$

Thus, the differential equation must *not* be exact.

**9** **What's the solution to this exact differential equation?**

$$3x^2 + 2y \frac{dy}{dx} = 0$$

**Solution:** $y = (c - x^3)^{1/2}$

1. From the original equation, you can identify either

$$M(x, y) = 3x^2$$

or

$$\frac{\partial f(x, y)}{\partial x} = 3x^2$$

2. Integrating gives you

$$f(x, y) = x^3 + g(y)$$

where $g(y)$ is a function.

3. From the original equation, you can also determine that

$$N(x, y) = 2y$$

Because

$$\frac{\partial f(x, y)}{\partial y} = N(x, y)$$

that means

$$\frac{\partial f(x, y)}{\partial y} = 2y$$

4. Also, because

$$f(x, y) = x^3 + g(y)$$

then

$$\frac{\partial f(x, y)}{\partial y} = \frac{\partial g(y)}{\partial y}$$

so

$$\frac{\partial g(y)}{\partial y} = 2y$$

5. Integrating gives you

$$g(y) = y^2 + k$$

where $k$ is a constant of integration.

6. Knowing that

$$f(x, y) = x^3 + g(y)$$

you get

$$f(x, y) = x^3 + y^2 + k$$

7. In general, the solution is

$$f(x, y) = c$$

so

$$c = x^3 + y^2 + k$$

8. Absorbing $k$ into $c$ gives you the following:

$$c = x^3 + y^2$$

9. Finally, solving for $y$ leaves you with

$$y = (c - x^3)^{1/2}$$

**10**  **Solve this exact differential equation:**

$$\frac{1}{y^2} - \frac{2x}{y^3} \frac{dy}{dx} = 0$$

**Solution: $y = (cx)^{1/2}$**

1. Here's what you know right off the bat:

$$M(x, y) = \frac{1}{y^2}$$

or

$$\frac{\partial f(x, y)}{\partial x} = \frac{1}{y^2}$$

2. Integrate to get the following:

$$f(x, y) = -\frac{1}{y} + g(y)$$

where $g(y)$ is a function of $y$.

3. Next up, identify that

$$N(x, y) = -\frac{2x}{y^3}$$

Because

$$\frac{\partial f(x, y)}{\partial y} = N(x, y)$$

you know that

$$\frac{\partial f(x, y)}{\partial y} = -\frac{2x}{y^3}$$

4. Because

$$f(x, y) = -\frac{1}{y} + g(y)$$

you can differentiate to get

$$\frac{\partial f(x, y)}{\partial y} = \frac{1}{y^2} + \frac{\partial g(y)}{\partial y}$$

so

$$\frac{\partial g(y)}{\partial y} = -\frac{1}{y^2} - \frac{2x}{y^3}$$

5. Integrate:

$$g(y) = \frac{1}{y} - \frac{x}{y^2} + k$$

where $k$ is a constant of integration.

6. Because

$$f(x, y) = -\frac{1}{y} + g(y)$$

you get this:

$$f(x, y) = -\frac{x}{y^2} + k$$

7. In general, the solution is

$$f(x, y) = m$$

where $m$ is a constant, so

$$m = -\frac{x}{y^2} + k$$

8. Absorb $k$ and the minus sign into $m$:

$$m = \frac{x}{y^2}$$

9. Then solve for $y$ (where $c = {}^1/_m$):

$$y = (cx)^{1/2}$$

**11** **What's the solution to this exact differential equation?**

$$y + 2x + x\,\frac{dy}{dx} = 0$$

**Solution:** $y = \dfrac{\left(c - x^2\right)}{x}$

1. From the original equation, you can identify either

$$M(x, y) = y + 2x$$

or

$$\frac{\partial f(x, y)}{\partial x} = y + 2x$$

2. Integrating gives you

$$f(x, y) = xy + x^2 + g(y)$$

where $g(y)$ is a function of $y$.

3. From the original equation, you can also determine that

$$N(x, y) = x$$

Because

$$\frac{\partial f(x, y)}{\partial y} = N(x, y)$$

that means

$$\frac{\partial f(x, y)}{\partial y} = x$$

4. Also, because

$$f(x, y) = xy + x^2 + g(y)$$

then differentiating with respect to $y$ should give you this:

$$\frac{\partial f(x, y)}{\partial y} = x + \frac{\partial g(y)}{\partial y}$$

so

$$\frac{\partial g(y)}{\partial y} = 0$$

5. Integrating gives you

$$g(y) = k$$

where $k$ is a constant of integration.

6. Knowing that

$$f(x, y) = xy + x^2 + g(y)$$

you get

$$f(x, y) = xy + x^2 + k$$

7. In general, the solution is

$$f(x, y) = c$$

where $c$ is a constant, so

$$c = xy + x^2 + k$$

8. Absorbing $k$ into $c$ gives you the following:

$$c = xy + x^2$$

9. Finally, solving for $y$ leaves you with

$$y = \frac{(c - x^2)}{x}$$

**12** **Solve this exact differential equation:**

$$\frac{1}{y} - \frac{x}{y^2} \frac{dy}{dx} = 0$$

**Solution: $y = \dfrac{x}{c}$**

1. Here's what you know right off the bat:

$$M(x, y) = \frac{1}{y}$$

or

$$\frac{\partial f(x, y)}{\partial x} = \frac{1}{y}$$

2. Integrate with respect to $x$ to get the following:

$$f(x, y) = \frac{x}{y} + g(y)$$

where $g(y)$ is a function of $y$.

3. Next up, identify that

$$N(x, y) = -\frac{x}{y^2}$$

Because

$$\frac{\partial f(x, y)}{\partial y} = N(x, y)$$

you know that

$$\frac{\partial f(x, y)}{\partial y} = -\frac{x}{y^2}$$

4. Because

$$f(x, y) = \frac{x}{y} + g(y)$$

you can differentiate with respect to $y$ to get

$$\frac{\partial f(x, y)}{\partial y} = -\frac{x}{y^2} + \frac{\partial g(y)}{\partial y}$$

so

$$\frac{\partial g(y)}{\partial y} = 0$$

5. Integrate:

$$g(y) = k$$

where $k$ is a constant of integration.

6. Because

$$f(x, y) = \frac{x}{y} + g(y)$$

you get this:

$$f(x, y) = \frac{x}{y} + k$$

7. In general, the solution is

$$f(x, y) = c$$

where $c$ is a constant, so

$$c = \frac{x}{y} + k$$

8. Absorb $k$ into $c$:

$$c = \frac{x}{y}$$

9. Then solve for $y$:

$$y = \frac{x}{c}$$

# Part II
# Finding Solutions to Second and Higher Order Differential Equations

## The 5th Wave
### By Rich Tennant

©RICHTENNANT

"This guy writes an equation for over 20 minutes, and he has the nerve to say, 'Voilà?'"

## In this part . . .

**G**et ready to take your differential equations skills to the next level as you practice solving second (and higher) order differential equations. You also get reacquainted with dazzling and time-saving techniques, such as the method of undetermined coefficients.

# Chapter 4

# Working with Linear Second Order Differential Equations

## In This Chapter

▶ Solving second order differential equations that are both linear and homogeneous

▶ Reveling in the three roots: real, distinct roots; complex roots; and real, identical roots

$A$re linear second order differential equations keeping you up at night? You know the ones. They look like this:

$y'' + p(x)y' + q(x)y = g(x)$

where

$$y'' = \frac{d^2y}{dx^2}$$

and

$$y' = \frac{dy}{dx}$$

Then this is the chapter for you because it's all about practicing solving linear second order differential equations that are also homogeneous.

***Note:*** In this chapter, you only solve differential equations for regions where $p(x)$, $q(x)$, and $g(x)$ are continuous functions (that is, where they don't take quick jumps in value). Usually there will be an initial condition, such as

$y(x_o) = y_0$

and

$y'(x_o) = y_0'$

# Getting the Goods on Linear Second Order Differential Equations

In linear second order differential equations, the exponent of $y''$, $y'$, and $y$ is 1. Equations not in that form are called *nonlinear* (but don't worry, I don't deal with those nasty things here).

The equation I present at the beginning of this chapter is the typical form used for linear second order differential equations in most textbooks. However, some textbooks seem to just want to make your life difficult, so they write the equation as follows:

$$P(x)y'' + Q(x)y' + R(x)y = G(x)$$

Note that the only difference in this form is that $y''$ has a coefficient, $P(x)$.

You can convert such equations into the first format I present simply by noting that

$$p(x) = \frac{Q(x)}{P(x)}$$

and

$$q(x) = \frac{R(x)}{P(x)}$$

and

$$g(x) = \frac{G(x)}{P(x)}$$

Of course, linear second order differential equations can also be *homogeneous,* meaning that in an equation such as the following, $g(x) = 0$:

$$y'' + p(x)y' + q(x)y = 0$$

Using the $P(x)$, $Q(x)$, $R(x)$, and $G(x)$ terminology that some textbooks prefer, note that you can also rewrite this equation with $G(x) = 0$:

$$P(x)y'' + Q(x)y' + R(x)y = 0$$

If a linear second order differential equation can't be put into either form, the equation is considered *nonhomogeneous.*

So now you're up to speed on what makes a second order differential equation both linear and homogeneous, but can you tell just by looking at an equation that it has both characteristics? Test yourself with the following practice problems.

**Q.** Is this differential equation linear and homogeneous?

$$\frac{d^2y}{dx^2} + 4\frac{dy}{dx} + 4y^2 = 0$$

**A.** Homogeneous but not linear

1. If you group all the nonconstant terms on the left side of the equation, the result equals 0, so you can cast the differential equation in this form (where the function $f(\ )$ has no constant terms):

$$\frac{d^2y}{dx^2} - f\left(x,\ y,\ \frac{dy}{dx}\right) = 0$$

Therefore, the differential equation is homogeneous.

2. You can't, however, put the differential equation into this form, because the exponent of $y$ is 2 (not 1):

$$y'' + p(x)y' + q(x)y = g(x)$$

Consequently, the differential equation isn't linear.

---

**1.** Is this differential equation linear and homogeneous?

$$\frac{d^2y}{dx^2} + 9\frac{dy}{dx} + 9y = 0$$

*Solve It*

**2.** Is the following equation both linear and homogeneous?

$$(y'')^2 + 4y' + 8y = 0$$

*Solve It*

**3.** Is this differential equation linear and homogeneous?

$$\frac{d^2y}{dx^2} + 9\frac{dy}{dx} + 9y = 5$$

Solve It

**4.** Is the following equation both linear and homogeneous?

$$\frac{d^2y}{dx^2} + 5\frac{dy}{dx} + 9y^2 = 3$$

Solve It

# Finding the Solution When Constant Coefficients Come into Play

After you know that the second order differential equation you're working with is both linear and homogeneous, the next step is to work it out. This section offers you some practice doing just that.

Following is an example of a second order differential equation that's both linear and homogeneous:

$$y'' - y = 0$$

where

$$y(0) = 9$$

and

$$y'(0) = -1$$

To solve this differential equation, you need a solution $y = f(x)$ whose second derivative is the same as $f(x)$ itself because subtracting the $f(x)$ from $f''(x)$ gives you 0. I bet you can think of one such solution: $y = e^x$. Substituting $y = e^x$ gives you

$$e^x - e^x = 0$$

so $y = e^x$ is a solution.

In fact, $y = c_1 e^x$ is also a solution, because $y''$ still equals $c_1 e^x$, which means that substituting $y = c_1 e^x$ gives you

$$c_1 e^x - c_1 e^x = 0$$

Guess what? That means $y = c_1 e^x$ is also a solution. In fact, it's more general than just $y = e^x$, because $y = c_1 e^x$ represents an infinite number of solutions, depending on the value of $c_1$.

You can keep on going if you note that $y = e^{-x}$ is also a solution because

$$y'' - y = e^{-x} - e^{-x} = 0$$

Of course, that realization alerts you to the fact that $y = c_2 e^{-x}$ is yet another solution because

$$y'' - y = c_2 e^{-x} - c_2 e^{-x} = 0$$

Hmmm. If $y = c_1 e^x$ is a solution and $y = c_2 e^{-x}$ is a solution, then the *sum* of these two solutions must also be a solution:

$$y = c_1 e^x + c_2 e^{-x}$$

To match the initial conditions, you can use the form of the solution $y = c_1 e^x + c_2 e^{-x}$, which means that $y' = c_1 e^x - c_2 e^{-x}$. Using the initial conditions, you get

$$y(0) = c_1 e^x + c_2 e^{-x} = c_1 + c_2 = 9$$
$$y'(0) = c_1 e^x - c_2 e^{-x} = c_1 - c_2 = -1$$

which leaves you with these two equations:

$$c_1 + c_2 = 9$$
$$c_1 - c_2 = -1$$

So now you have two equations in two unknowns. To solve them, write the first equation in this form:

$$c_2 = 9 - c_1$$

Then substitute this expression for $c_2$ into the other equation:

$$c_1 - 9 + c_1 = -1$$

or

$$2c_1 = 8$$

so

$$c_1 = 4$$

Substituting this value of $c_1$ into the first equation ($c_1 + c_2 = 9$) gives you

$$4 + c_2 = 9$$

or

$$c_2 = 5$$

The values of $c_1$ and $c_2$ give you your solution, which is

$$y = 4e^x + 5e^{-x}$$

## Rooted in reality: Second order differential equations with real and distinct roots

Guessing solutions is great when it works, but you can't count on always being lucky enough to get your desired answer. Instead, try assuming a solution of the form

$$y = e^{rx}$$

and plug that into the differential equation you're working on. In the case of $y'' - y = 0$, you get

$$r^2 y - y = 0$$

Dividing by $y$ gives you

$$r^2 - 1 = 0$$

which is actually the *characteristic equation* (the equation you get when you substitute in your assumed solution) for the differential equation. After you find the roots of the characteristic equation, $r_1$ and $r_2$, you can determine that the solution to the differential equation is

$$y = c_1 e^{r_1 x} + c_2 e^{r_2 x}$$

REMEMBER

Three types of solutions are possible for the characteristic equation:

- $r_1$ and $r_2$ are real and distinct
- $r_1$ and $r_2$ are complex numbers (complex conjugates of each other)
- $r_1 = r_2$ where $r_1$ and $r_2$ are real

Here's an example problem that shows you how to solve a differential equation in which $r_1$ and $r_2$ are real and distinct. Check it out and then try to solve the practice problems on your own.

EXAMPLE

**Q.** Solve this differential equation:

$$\frac{d^2y}{dx^2} + 3\frac{dy}{dx} + 2y = 0$$

where

$$y(0) = 2$$

$$y'(0) = -3$$

**A.** $y = e^{-x} + e^{-2x}$

1. Assume a solution of the form

$$y = e^{rx}$$

2. Plug the assumed solution into the differential equation to get

$$r^2y + 3ry + 2y = 0$$

3. Divide by $y$ to get the characteristic equation:

$$r^2 + 3r + 2 = 0$$

4. Use the quadratic equation to get the two roots of the characteristic equation:

$$r_1 = -1 \text{ and } r_2 = -2$$

5. Plug the two roots in to get the following general solution:

$$y = c_1e^{-x} + c_2e^{-2x}$$

6. Find the derivative, $y'$:

$$y' = -c_1e^{-x} - 2c_2e^{-2x}$$

7. Use the initial conditions to get the first equation

$$y(0) = c_1 + c_2 = 2$$

and the second equation

$$y'(0) = -c_1 - 2c_2 = -3$$

8. Add the first and second equations together to get

$$-c_2 = -1 \text{ or } c_2 = 1$$

9. Then plug this result into the first equation:

$$c_1 + 1 = 2, \text{ or } c_1 = 1$$

10. Finally, use $c_1$ and $c_2$ to get the solution:

$$y = e^{-x} + e^{-2x}$$

**5.** Solve this differential equation:

$$\frac{d^2y}{dx^2} + 4\frac{dy}{dx} + 3y = 0$$

where

$y(0) = 2$

$y'(0) = -4$

*Solve It*

**6.** What's the solution to this differential equation?

$$\frac{d^2y}{dx^2} + 5\frac{dy}{dx} + 6y = 0$$

where

$y(0) = 2$

$y'(0) = -5$

*Solve It*

**7.** Solve this differential equation:

$$\frac{d^2y}{dx^2} + 5\frac{dy}{dx} + 6y = 0$$

where

$y(0) = 3$

$y'(0) = -8$

*Solve It*

**8.** What's the solution to this differential equation?

$$\frac{d^2y}{dx^2} + 6\frac{dy}{dx} + 8y = 0$$

where

$y(0) = 4$

$y'(0) = -14$

*Solve It*

**9.** Solve this differential equation:

$$\frac{d^2y}{dx^2} + 6\frac{dy}{dx} + 5y = 0$$

where

$y(0) = 3$

$y'(0) = -11$

*Solve It*

**10.** What's the solution to this differential equation?

$$\frac{d^2y}{dx^2} + 7\frac{dy}{dx} + 6y = 0$$

where

$y(0) = 5$

$y'(0) = -25$

*Solve It*

# Adding complexity: Second order differential equations with complex roots

Now it's time to solve second order differential equations where the roots of the characteristic equation are *complex,* meaning they involve the imaginary number $i$. In this case, the roots, which you get from the quadratic equation, are of the form

$r_1 = m + in$ and $r_2 = m - in$

where $m$ and $n$ are both real numbers.

The solutions to the differential equation are

$y_1 = e^{r_1 x}$ and $y_2 = e^{r_2 x}$

so

$y_1 = e^{(m + in)x}$

and

$y_2 = e^{(m - in)x}$

Time to resort to two well-known formulas:

$$e^{iax} = \cos ax + i \sin ax$$

and

$$e^{-iax} = \cos ax - i \sin ax$$

to get the following:

$$y_1 = e^{(m + in)x} = e^{mx}(\cos nx + i \sin nx)$$

and

$$y_2 = e^{(m - in)x} \doteq e^{mx}(\cos nx - i \sin nx)$$

By adding these two solutions and doing a little trigonometry, you can cast the solution as

$$y(x) = c_1 e^{mx} \cos nx + c_2 e^{mx} \sin nx$$

Want to walk through the process again? Review the following example. Or if you're feeling up to the challenge, skip ahead to the practice problems.

**EXAMPLE**

**Q.** Solve this differential equation:

$$2y'' + 2y' + y = 0$$

where

$$y(0) = 1$$

and

$$y'(0) = 0$$

**A.** $y(x) = e^{-x/2} \cos\left(\dfrac{x}{2}\right) + e^{-x/2} \sin\left(\dfrac{x}{2}\right)$

1. Get the following characteristic equation:

   $$2r^2 + 2r + 1 = 0$$

2. Use the quadratic equation to solve for the two roots:

   $$r_1 = -\tfrac{1}{2} + (\tfrac{1}{2})i$$

   and

   $$r_2 = -\tfrac{1}{2} - (\tfrac{1}{2})i$$

3. Put the solutions in this form:

   $$r_1 = m + in \text{ and } r_2 = m - in$$

In this case,

$$m = -\tfrac{1}{2} \text{ and } n = \tfrac{1}{2}$$

4. Use the equation:

   $$y(x) = c_1 e^{mx} \cos nx + c_2 e^{mx} \sin nx$$

5. So the solution is

   $$y(x) = c_1 e^{-x/2} \cos\left(\dfrac{x}{2}\right) + c_2 e^{-x/2} \sin\left(\dfrac{x}{2}\right)$$

6. Use the initial conditions to find $c_1$ and $c_2$. Plugging into the initial conditions gives you the following first equation:

   $$y(0) = c_1 = 1$$

   as well as this second one:

   $$y'(0) = -\dfrac{c_1}{2} + \dfrac{c_2}{2} = 0$$

7. Substituting $c_1 = 1$ into the second equation gives you

   $$y'(0) = -\dfrac{1}{2} + \dfrac{c_2}{2} = 0$$

   so $c_2 = 1$, which makes the solution

   $$y(x) = e^{-x/2} \cos\left(\dfrac{x}{2}\right) + e^{-x/2} \sin\left(\dfrac{x}{2}\right)$$

**11.** Solve this differential equation:

$$y'' + 4y' + 5y = 0$$

where

$$y(0) = 1$$

and

$$y'(0) = -1$$

*Solve It*

**12.** Find the solution to this differential equation:

$$y'' + 4y = 0$$

where

$$y(0) = 1$$

and

$$y'(0) = 1$$

*Solve It*

# Look-alike city: Second order differential equations with real, identical roots

The third and final type of solution for a characteristic equation involves identical real roots. If you substitute $y = e^{rx}$ into the following differential equation:

$$ay'' + by' + cy = 0$$

you get this:

$$ar^2 + br + c = 0$$

For the purposes of this section, the roots are

$$r_1 = {}^{-b/2a} \text{ and } r_2 = {}^{-b/2a}$$

That's a problem, because you get these two solutions (which differ only by a constant):

$$y_1 = c_1 e^{-bx/2a} \text{ and } y_2 = c_2 e^{-bx/2a}$$

Because the difference between these two equations is merely a constant, they're really the same solution. Ah, but the fun doesn't stop there! The actual two solutions really differ by a factor of $x$ (see the proof for yourself in *Differential Equations For Dummies*), so the real solution is

$$y(x) = c_1 x e^{-bx/2a} + c_2 e^{-bx/2a}$$

Following is another example so you can see the process in action before trying to solve the next four practice problems on your own.

**Q.** Solve this differential equation:

$$y'' + 2y' + y = 0$$

where

$$y(0) = 1$$

and

$$y'(0) = 1$$

**A.** $y(x) = 2xe^{-x} + e^{-x}$

1. Solve for the characteristic equation:

$$r^2 + 2r + 1 = 0$$

2. Then factor the characteristic equation this way:

$$(r + 1)(r + 1) = 0$$

Looks like the roots of the differential equation are identical, $-1$ and $-1$.

3. Therefore, the solution is of the form

$$y(x) = c_1 x e^{-x} + c_2 e^{-x}$$

4. To find $c_1$ and $c_2$, use the initial conditions. Substitute the first initial condition into the solution to get

$$y(0) = c_2 = 1$$

So $c_2 = 1$.

5. Differentiate the solution to get $y'(x)$:

$$y'(x) = c_1 e^{-x} - c_1 x e^{-x} = c_1 e^{-x}(1 - x) - c_2 e^{-x}$$

6. From the initial condition for $y'(0)$, you get

$$y'(0) = c_1 - 1 = 1$$

So $c_1 = 2$.

7. Plugging in $c_1$ and $c_2$ gives you the following general solution:

$$y(x) = 2xe^{-x} + e^{-x}$$

**13.** Solve this differential equation:

$$y'' + 4y' + 4y = 0$$

where

$$y(0) = 1$$

and

$$y'(0) = 0$$

*Solve It*

**14.** What's the solution to this equation?

$$y'' + 10y' + 25y = 0$$

where

$$y(0) = 1$$

and

$$y'(0) = 2$$

*Solve It*

**15.** Solve this differential equation:

$$y'' + 8y' + 16y = 0$$

where

$$y(0) = 2$$

and

$$y'(0) = 4$$

*Solve It*

**16.** What's the solution to this equation?

$$y'' + 6y' + 9y = 0$$

where

$$y(0) = 4$$

and

$$y'(0) = 4$$

*Solve It*

# Answers to Linear Second Order Differential Equation Problems

Here are the answers to the practice questions I provide throughout this chapter. I walk you through each answer so you can see the problems worked out step by step. Enjoy!

**1** **Is this differential equation linear and homogeneous?**

$$\frac{d^2y}{dx^2} + 9\frac{dy}{dx} + 9y = 0$$

**Answer: Homogeneous and linear**

1. If you group all the nonconstant terms on the left side, the result equals 0, so you can cast the differential equation in this form (where the function $f(\ )$ has no constant terms):

$$\frac{d^2y}{dx^2} - f\left(x,\ y,\ \frac{dy}{dx}\right) = 0$$

Therefore, the differential equation is homogeneous.

2. You can also put the differential equation into this form, because the exponent of $y$ is 2:

$$y'' + p(x)y' + q(x)y = g(x)$$

Consequently, the differential equation is linear.

**2** **Is the following equation both linear and homogeneous?**

$$(y'')^2 + 4y' + 8y = 0$$

**Answer: Homogeneous but not linear**

1. Because all nonconstant terms can be grouped on the left side of the equation, your result equals 0. Consequently, you can cast the differential equation in this form (where the function $f(\ )$ has no constant terms):

$$(y'')^2 - f(x, y, y') = 0$$

This differential equation is homogeneous.

2. You can't put the differential equation into this form, because the exponent of $y''$ is 2 (not 1):

$$y'' + p(x)y' + q(x)y = g(x)$$

This equation isn't linear.

**3** **Is this differential equation linear and homogeneous?**

$$\frac{d^2y}{dx^2} + 9\frac{dy}{dx} + 9y = 5$$

**Answer: Nonhomogeneous but linear**

1. In this case, you can't group all the nonconstant terms on the left side of the equation and have the result equal 0, so you can't cast the differential equation in this form (where the function $f(\ )$ has no constant terms):

$$\frac{d^2y}{dx^2} - f\left(x,\ y,\ \frac{dy}{dx}\right) = 0$$

Therefore, the differential equation isn't homogeneous.

2. You can put the differential equation into this form, because the exponent of $y$ is 2:

$$y'' + p(x)y' + q(x)y = g(x)$$

Consequently, the differential equation is linear.

**4** **Is the following equation both linear and homogeneous?**

$$\frac{d^2y}{dx^2} + 5\frac{dy}{dx} + 9y^2 = 3$$

**Answer: Neither homogeneous nor linear**

1. Not all the nonconstant terms can be grouped on the left side and have the result equal 0, so you can't cast the differential equation in this form (where the function $f(\,)$ has no constant terms):

$$\frac{d^2y}{dx^2} - f\left(x,\ y,\ \frac{dy}{dx}\right) = 0$$

This differential equation isn't homogeneous.

2. You can't put the differential equation into this form, because the exponent of $y''$ is 2 (not 1):

$$y'' + p(x)y' + q(x)y = g(x)$$

This equation isn't linear.

**5** **Solve this differential equation:**

$$\frac{d^2y}{dx^2} + 4\frac{dy}{dx} + 3y = 0$$

**where**

$$y(0) = 2$$

$$y'(0) = -4$$

**Solution: $y = e^{-x} + e^{-3x}$**

1. Assume a solution of the form

$$y = e^{rx}$$

2. Plug the assumed solution into the differential equation to get

$$r^2y + 4ry + 3y = 0$$

3. Divide by $y$ to find the characteristic equation:

$$r^2 + 4r + 3 = 0$$

4. Use the quadratic equation to get the two roots of the characteristic equation:

$$r_1 = -1 \text{ and } r_2 = -3$$

5. Plug the two roots in to get the general solution:

$$y = c_1e^{-x} + c_2e^{-3x}$$

6. Next, find the derivative, $y'$:

$$y' = -c_1e^{-x} - 3c_2e^{-3x}$$

7. Use the initial conditions to get the first equation:

$$y(0) = c_1 + c_2 = 2$$

and the second equation:

$$y'(0) = -c_1 - 3c_2 = -4$$

8. Add the first and second equations together to get

$$-2c_2 = -2, \text{ or } c_2 = 1$$

9. Plug this result into the first equation:

$$c_1 + 1 = 2, \text{ or } c_1 = 1$$

10. Then just use $c_1$ and $c_2$ to get your solution:

$$y = e^{-x} + e^{-3x}$$

**6** **What's the solution to this differential equation?**

$$\frac{d^2y}{dx^2} + 5\frac{dy}{dx} + 6y = 0$$

**where**

$$y(0) = 2$$

$$y'(0) = -5$$

**Solution: $y = e^{-2x} + e^{-3x}$**

1. Assume a solution of the form

$$y = e^{rx}$$

2. Plug that solution into the differential equation:

$$r^2y + 5ry + 6y = 0$$

3. Then divide by $y$ to get the characteristic equation:

$$r^2 + 5r + 6 = 0$$

4. Find the two roots of the characteristic equation by using the quadratic equation:

$$r_1 = -2 \text{ and } r_2 = -3$$

5. Plug the two roots in to get your general solution:

$$y = c_1e^{-2x} + c_2e^{-3x}$$

6. Then find the derivative, $y'$:

$$y' = -2c_1e^{-2x} - 3c_2e^{-3x}$$

7. Use the initial conditions to get the first equation:

$$y(0) = c_1 + c_2 = 2$$

and the second equation:

$$y'(0) = -2c_1 - 3c_2 = -5$$

8. Then add twice the first equation to the second equation:

$$-c_2 = -1, \text{ or } c_2 = 1$$

9. Plug this result into the first equation to get

$c_1 + 1 = 2$, or $c_1 = 1$

10. Finally, use $c_1$ and $c_2$ to find your solution:

$y = e^{-2x} + e^{-3x}$

**7** **Solve this differential equation:**

$$\frac{d^2y}{dx^2} + 5\frac{dy}{dx} + 6y = 0$$

**where**

$y(0) = 3$

$y'(0) = -8$

**Solution:** $y = e^{-2x} + 2e^{-3x}$

1. Assume a solution of the form

$y = e^{rx}$

2. Plug the assumed solution into the differential equation to get

$r^2y + 5ry + 6y = 0$

3. Divide by $y$ to find the characteristic equation:

$r^2 + 5r + 6 = 0$

4. Use the quadratic equation to get the two roots of the characteristic equation:

$r_1 = -2$ and $r_2 = -3$

5. Plug the two roots in to get the general solution:

$y = c_1e^{-2x} + c_2e^{-3x}$

6. Next, find the derivative, $y'$:

$y' = -2c_1e^{-2x} - 3c_2e^{-3x}$

7. Use the initial conditions to get the first equation:

$y(0) = c_1 + c_2 = 3$

and the second equation:

$y'(0) = -2c_1 - 3c_2 = -8$

8. Add twice the first equation to the second equation to get

$-c_2 = -2$, or $c_2 = 2$

9. Plug this result into the first equation:

$c_1 + 2 = 3$, or $c_1 = 1$

10. Then just use $c_1$ and $c_2$ to get your solution:

$y = e^{-2x} + 2e^{-3x}$

**8** **What's the solution to this differential equation?**

$$\frac{d^2y}{dx^2} + 6\frac{dy}{dx} + 8y = 0$$

where

$y(0) = 4$

$y'(0) = -14$

**Solution:** $y = e^{-2x} + 3e^{-4x}$

1. Assume a solution of the form

$y = e^{rx}$

2. Plug that solution into the differential equation:

$r^2 y + 6ry + 8y = 0$

3. Then divide by $y$ to get the characteristic equation:

$r^2 + 6r + 8 = 0$

4. Find the two roots of the characteristic equation by using the quadratic equation:

$r_1 = -2$ and $r_2 = -4$

5. Plug the two roots in to get your general solution:

$y = c_1 e^{-2x} + c_2 e^{-4x}$

6. Then find the derivative, $y'$:

$y' = -2c_1 e^{-2x} - 4c_2 e^{-4x}$

7. Use the initial conditions to get the first equation:

$y(0) = c_1 + c_2 = 4$

and the second equation:

$y'(0) = -2c_1 - 4c_2 = -14$

8. Then add twice the first equation to the second equation:

$-2c_2 = -6$, or $c_2 = 3$

9. Plug this result into the first equation to get

$c_1 + 3 = 4$, or $c_1 = 1$

10. Finally, use $c_1$ and $c_2$ to find your solution:

$y = e^{-2x} + 3e^{-4x}$

**9** **Solve this differential equation:**

$$\frac{d^2 y}{dx^2} + 6\frac{dy}{dx} + 5y = 0$$

where

$y(0) = 3$

$y'(0) = -11$

**Solution:** $y = e^{-x} + 2e^{-5x}$

1. Assume a solution of the form

$y = e^{rx}$

2. Plug the assumed solution into the differential equation to get

   $r^2 y + 6ry + 5y = 0$

3. Divide by $y$ to find the characteristic equation:

   $r^2 + 6r + 5 = 0$

4. Use the quadratic equation to get the two roots of the characteristic equation:

   $r_1 = -1$ and $r_2 = -5$

5. Plug the two roots in to get the general solution:

   $y = c_1 e^{-x} + c_2 e^{-5x}$

6. Next, find the derivative, $y'$:

   $y' = -c_1 e^{-x} - 5c_2 e^{-5x}$

7. Use the initial conditions to get the first equation:

   $y(0) = c_1 + c_2 = 3$

   and the second equation:

   $y'(0) = -c_1 - 5c_2 = -11$

8. Add the first equation to the second equation to get

   $-4c_2 = -8$, or $c_2 = 2$

9. Plug this result into the first equation:

   $c_1 + 2 = 3$, or $c_1 = 1$

10. Then just use $c_1$ and $c_2$ to get your solution:

    $y = e^{-x} + 2e^{-5x}$

**10** **What's the solution to this differential equation?**

$$\frac{d^2 y}{dx^2} + 7\frac{dy}{dx} + 6y = 0$$

**where**

**$y(0) = 5$**

**$y'(0) = -25$**

**Solution: $y = e^{-x} + 4e^{-6x}$**

1. Assume a solution of the form

   $y = e^{rx}$

2. Plug that solution into the differential equation:

   $r^2 y + 7ry + 6y = 0$

3. Then divide by $y$ to get the characteristic equation:

   $r^2 + 7r + 6 = 0$

4. Find the two roots of the characteristic equation by using the quadratic equation:

   $r_1 = -1$ and $r_2 = -6$

5. Plug the two roots in to get the general solution:

$$y = c_1 e^{-x} + c_2 e^{-6x}$$

6. Then find the derivative, $y'$:

$$y' = -c_1 e^{-x} - 6c_2 e^{-6x}$$

7. Use the initial conditions to get the first equation:

$$y(0) = c_1 + c_2 = 5$$

and the second equation:

$$y'(0) = -c_1 - 6c_2 = -25$$

8. Then add the first equation to the second equation:

$$-5c_2 = -20, \text{ or } c_2 = 4$$

9. Plug this result into the first equation to get

$$c_1 + 4 = 5, \text{ or } c_1 = 1$$

10. Finally, use $c_1$ and $c_2$ to find your solution:

$$y = e^{-x} + 4e^{-6x}$$

**11** **Solve this differential equation:**

$$y'' + 4y' + 5y = 0$$

**where**

$$y(0) = 1$$

**and**

$$y'(0) = -1$$

**Solution: $y(x) = e^{-2x} \cos(x) + e^{-2x} \sin(x)$**

1. Get the following characteristic equation:

$$r^2 + 4 = 0$$

2. Solve for the two roots:

$$r_1 = 2i \text{ and } r_2 = -2i$$

3. Put the solutions into these forms:

$$r_1 = m + in \text{ and } r_2 = m - in$$

In this case,

$$m = 0 \text{ and } n = 2$$

4. Use the equation:

$$y(x) = c_1 e^{mx} \cos nx + c_2 e^{mx} \sin nx$$

5. So the solution is

$$y(x) = c_1 \cos(2x) + c_2 \sin(2x)$$

6. Use the initial conditions to find $c_1$ and $c_2$. Plugging into the initial conditions gives you the following first equation:

$$y(0) = c_1 = 1$$

as well as this second one:

$$y'(0) = 2c_1 = 4$$

7. Substituting $c_1 = 1$ into the second equation gives you

$$y'(0) = -2 + c_2 = -1$$

so $c_2 = 1$.

8. Plug $c_1$ and $c_2$ in to find the solution:

$$y(x) = e^{-2x} \cos(x) + e^{-2x} \sin(x)$$

**12**  **Find the solution to this differential equation:**

$$y'' + 4y = 0$$

**where**

$$y(0) = 1$$

**and**

$$y'(0) = 1$$

**Solution:** $y(x) = \cos(2x) + \dfrac{1}{2} \sin(2x)$

1. Find the characteristic equation:

$$r^2 + 4 = 0$$

2. Then solve for the two roots:

$$r_1 = 2i \text{ and } r_2 = -2i$$

3. Put the solutions in the following forms:

$$r_1 = m + in \text{ and } r_2 = m - in$$

In this case,

$$m = 0 \text{ and } n = 2$$

4. Use this equation:

$$y(x) = c_1 e^{mx} \cos nx + c_2 e^{mx} \sin nx$$

to get this solution:

$$y(x) = c_1 \cos(2x) + c_2 \sin(2x)$$

5. Use the initial conditions to find $c_1$ and $c_2$; plug into the initial conditions to get the first equation:

$$y(0) = c_1 = 1$$

and the second one:

$$y'(0) = 2c_2 = 1$$

6. You now know that $c_1 = 1$. Go ahead and solve for $c_2$:

$$y'(0) = 2c_2 = \dfrac{1}{2}$$

7. Plugging $c_1$ and $c_2$ in gives you your solution, which is

$$y(x) = \cos(2x) + \dfrac{1}{2} \sin(2x)$$

**13** **Solve this differential equation:**

$$y'' + 4y' + 4y = 0$$

**where**

$$y(0) = 1$$

**and**

$$y'(0) = 0$$

**Solution:** $y(x) = 2xe^{-2x} + e^{-2x}$

1. Solve for the characteristic equation:

$$r^2 + 4r + 4 = 0$$

2. Then factor the characteristic equation this way:

$$(r + 2)(r + 2) = 0$$

You now know that the roots of the differential equation are identical, –2 and –2.

3. Therefore, the solution is of the form

$$y(x) = c_1 xe^{-2x} + c_2 e^{-2x}$$

4. To find $c_1$ and $c_2$, use the initial conditions provided and substitute the first initial condition into the solution to get

$$y(0) = c_2 = 1$$

So $c_2 = 1$.

5. Differentiate the solution to get $y'(x)$:

$$y'(x) = c_1 e^{-2x} - 2c_1 xe^{-2x} - 2c_2 e^{-2x} = c_1 e^{-2x}(1 - 2x) - 2c_2 e^{-2x}$$

6. From the initial condition for $y'(0)$, you get

$$y'(0) = c_1 - 2c_2 = 0$$

So $c_1 = 2$.

7. Plugging in $c_1$ and $c_2$ gives you this general solution:

$$y(x) = 2xe^{-2x} + e^{-2x}$$

**14** **What's the solution to this equation?**

$$y'' + 10y' + 25y = 0$$

**where**

$$y(0) = 1$$

**and**

$$y'(0) = 2$$

**Solution:** $y(x) = 7xe^{-5x} + e^{-5x}$

1. Find the characteristic equation:

$$r^2 + 10r + 25 = 0$$

2. Factor it as follows:

$$(r + 5)(r + 5) = 0$$

Hmmm, the roots of the differential equation are identical, –5 and –5, so the solution must be of this form:

$$y(x) = c_1 x e^{-5x} + c_2 e^{-5x}$$

3. Use the initial conditions to find $c_1$ and $c_2$. Substituting the first initial condition into the solution gives you

$$y(0) = c_2 = 1$$

so $c_2 = 1$.

4. Differentiating the solution gives you $y'(x)$

$$y'(x) = c_1 e^{-5x} - 5c_1 x e^{-5x} - 5c_2 e^{-5x} = c_1 e^{-5x}(1 - 5x) - 5c_2 e^{-5x}$$

5. From the initial condition for $y'(0)$, you get

$$y'(0) = c_1 - 5 = 2$$

so $c_1 = 7$.

6. Plug $c_1$ and $c_2$ in to obtain the general solution:

$$y(x) = 7x e^{-5x} + e^{-5x}$$

**15** **Solve this differential equation:**

$$y'' + 8y' + 16y = 0$$

**where**

$$y(0) = 2$$

**and**

$$y'(0) = 4$$

**Solution:** $y(x) = 12x e^{-4x} + 2e^{-4x}$

1. Solve for the characteristic equation:

$$r^2 + 8r + 16 = 0$$

2. Then factor the characteristic equation this way:

$$(r + 4)(r + 4) = 0$$

You now know that the roots of the differential equation are identical, –4 and –4.

3. Therefore, the solution is of the form

$$y(x) = c_1 x e^{-4x} + c_2 e^{-4x}$$

4. To find $c_1$ and $c_2$, use the initial conditions provided and substitute the first initial condition into the solution to get

$$y(0) = c_2 = 2$$

So $c_2 = 2$.

5. Differentiate the solution to get $y'(x)$:

$$y'(x) = c_1 e^{-4x} - 4c_1 xe^{-4x} - 4c_2 e^{-4x} = c_1 e^{-4x}(1 - 4x) - 4c_2 e^{-4x}$$

6. From the initial condition for $y'(0)$, you get

$$y'(0) = c_1 - 4c_2 = c_1 - 8 = 4$$

So $c_1 = 12$.

7. Plugging in $c_1$ and $c_2$ gives you this general solution:

$$y(x) = 12xe^{-4x} + 2e^{-4x}$$

**16** **What's the solution to this equation?**

$$y'' + 6y' + 9y = 0$$

**where**

$$y(0) = 4$$

**and**

$$y'(0) = 4$$

**Solution:** $y(x) = 16xe^{-3x} + e^{-3x}$

1. Find the characteristic equation:

$$r^2 + 6r + 9 = 0$$

2. Factor it as follows:

$$(r + 3)(r + 3) = 0$$

Hmmm, the roots of the differential equation are identical, $-3$ and $-3$, so the solution must be of this form:

$$y(x) = c_1 xe^{-3x} + c_2 e^{-3x}$$

4. Use the initial conditions to find $c_1$ and $c_2$. Substituting the first initial condition into the solution gives you

$$y(0) = c_2 = 4$$

so $c_2 = 4$.

5. Differentiating the solution gives you $y'(x)$:

$$y'(x) = c_1 e^{-3x} - 3c_1 xe^{-3x} - 3c_2 e^{-3x} = c_1 e^{-3x}(1 - 3x) - 3c_2 e^{-3x}$$

6. From the initial conditions for $y'(0)$, you get

$$y'(0) = c_1 - 3 c_2 = c_1 - 12 = 4$$

so $c_1 = 16$.

7. Plug in $c_1$ and $c_2$ to obtain the general solution:

$$y(x) = 16xe^{-3x} + e^{-3x}$$

# Chapter 5

# Tackling Nonhomogeneous Linear Second Order Differential Equations

## In This Chapter
▶ Refreshing your memory of the method of undetermined coefficients
▶ Working with $g(x)$ in its various forms

**W**elcome to the wonderful world of nonhomogeneous second order differential equations! (If you're thinking "Homoge-huh?" flip to Chapter 4 for a refresher on what makes a second order differential equation homogeneous in the first place.) In other words, here's your chance to play with equations that look like this:

$$y'' + p(x)y' + q(x)y = g(x)$$

where $g(x) \neq 0$.

You, lucky person that you are, get to handle linear second order differential equations like this one in the following pages:

$$y'' - y' - 2y = 10e^4 x$$

The *method of undetermined coefficients* advises that when you find a candidate solution, $y$, and plug it into the left-hand side of the equation, you end up with $g(x)$. Because $g(x)$ is just a function of $x$, you can often guess the form of $y_p(x)$, up to arbitrary coefficients, and then solve for those coefficients by plugging $y_p(x)$ into the differential equation.

This method works because you're handling only $g(x)$, and the form of $g(x)$ can often tell you what a particular solution looks like. For example, if $g(x)$ is in the form of

- ▶ $e^{rx}$, then try a particular solution of the form $Ae^{rx}$, where $A$ is a constant. Because derivatives of $e^{rx}$ reproduce $e^{rx}$, you have a good chance of finding a particular solution this way.

- ▶ **a polynomial of order $n$,** then try a polynomial of order $n$. For instance, if $g(x) = x^2 + 1$, try a polynomial of the form $Ax^2 + Bx + C$.

- ▶ **a combination of sines and cosines,** $\sin \alpha x + \cos \beta x$, then try a combination of sines and cosines with undetermined coefficients, $A \sin \alpha x + B \cos \beta x$. Then plug into the differential equation and solve for $A$ and $B$.

You get practice with each of these problem types within this chapter as you work to find the general solution to each equation.

In order to find the general solution to a nonhomogeneous linear second order differential equation, you must add the corresponding homogeneous equation's solution to a particular solution of the nonhomogeneous equation (a *particular solution* is any solution of the non-homogeneous differential equation).

# Finding the General Solution for Differential Equations with a Nonhomogeneous $e^{rx}$ Term

As you may know from class or from *Differential Equations For Dummies,* you can't just find the solution to a nonhomogeneous linear second order differential equation that happens to give you $g(x)$ when you plug it in. You have to do some extra work by adding the solution to the homogeneous version of the same differential equation.

Think about it. Say you have this differential equation:

$$y'' + p(x)y' + q(x)y = g(x)$$

and the following solution gives you $g(x)$ when you plug it in:

$$y = y_p(x)$$

In order for your answer to be correct, you must add in the homogeneous solution; when you plug that into the differential equation, you get 0. So the general solution to the differential equation is

$$y = c_1 y_1(x) + c_2 y_2(x) + y_p(x)$$

where $c_1 y_1(x) + c_2 y_2(x)$ is the solution of the corresponding homogeneous differential equation:

$$y'' + p(x)y' + q(x)y = 0$$

That is, $y_1$ and $y_2$ are a fundamental set of solutions to the homogeneous differential equation, and $y_p(x)$ is a particular (or specific) solution to the nonhomogenuous equation.

So to solve a second order differential equation that's both linear *and* nonhomogeneous, you follow these overall steps:

1. **Find the corresponding homogeneous differential equation by setting $g(x)$ to 0.**

2. **Find the general solution, $y = c_1 y_1(x) + c_2 y_2(x)$, of the corresponding homogeneous differential equation.**

   This general solution of the homogeneous equation is referred to as $y_h$.

3. **Find a single solution to the nonhomogeneous equation.**

   This solution is sometimes referred to as the particular (or specific) solution, $y_p$.

4. **The general solution of the nonhomogeneous differential equation is the sum of $y_h + y_p$.**

The following example problem walks you through each of these steps. Take a few minutes to check it out and then try your hand at the practice problems.

**Q.** Find the general solution to this differential equation:

$$y'' - y' - 2y = 10e^{4x}$$

**A.** $y = c_1e^{-x} + c_2e^{2x} + e^{4x}$

1. Start by getting the homogeneous version of the differential equation:

$$y'' - y' - 2y = 0$$

2. Assume that the solution to the homogeneous differential equation is of the form $y = e^{rx}$. When you substitute that into the differential equation, you get this as your characteristic equation:

$$r^2 - r - 2 = 0$$

3. Factor the characteristic equation this way:

$$(r + 1)(r - 2) = 0$$

4. You can now determine that the roots, $r_1$ and $r_2$, of the characteristic equation are –1 and 2, giving you

$$y_1 = e^{-x} \text{ and } y_2 = e^{2x}$$

5. So the general solution to the homogeneous differential equation is given by

$$y = c_1e^{-x} + c_2e^{2x}$$

6. Now you need a particular solution to the nonhomogeneous differential equation that you started with:

$$y'' - y' - 2y = 10e^{4x}$$

Because $g(x)$ has the form $e^{4x}$ here, you can assume that the particular solution has the form

$$y_p(x) = Ae^{4x}$$

7. Substitute $y_p(x)$ into the differential equation to get

$$16Ae^{4x} - 4Ae^{4x} - 2Ae^{4x} = 10e^{4x}$$

8. Cancel out the $e^{4x}$ term:

$$16A - 4A - 2A = 10$$

or

$$10A = 10$$

so $A = 1$.

9. Tada! Your particular solution is

$$y_p(x) = e^{4x}$$

10. Now, because the general solution of the nonhomogeneous equation that you started with is the sum of the corresponding homogeneous equation's general solution and a particular solution of the nonhomogeneous equation, you get the following as your solution:

$$y = y_h + y_p$$

which is actually

$$y = c_1e^{-x} + c_2e^{2x} + e^{4x}$$

**1.** What's the general solution to this non-homogeneous second order differential equation?

$$y'' + 3y' + 2y = 6e^x$$

*Solve It*

**2.** Find the general solution to the following nonhomogeneous second order differential equation:

$$y'' + 4y' + 3y = 30e^{2x}$$

*Solve It*

**3.** Solve for the general solution to this equation:

$$y'' + 5y' + 6y = 36e^x$$

*Solve It*

**4.** What's the general solution to this non-homogeneous second order differential equation?

$$y'' + 2y' + y = 8e^x$$

*Solve It*

**5.** Find the general solution to the following nonhomogeneous second order differential equation:

$$y'' + 6y' + 8y = 70e^{3x}$$

Solve It

**6.** Solve for the general solution to this equation:

$$y'' + 6y' + 5y = 36e^x$$

Solve It

# Getting the General Solution When g (x) Is a Polynomial

Sometimes $g(x)$ acts as a polynomial, like in the case of $g(x) = ax^n + bx^{n-1} + cx^{n-2}$ (where $a$, $b$, and $c$ are all constants). You need to know how to handle such situations so you can find the general solution to the nonhomogeneous second order differential equation in question.

REMEMBER

Here's how to solve a differential equation of this form, $ay'' + by' + cy = g(x)$, where $a$, $b$, and $c$ are constants and $g(x)$ is a polynomial of order $n$:

1. **If $g(x)$ is a polynomial, you can assume the particular solution is of the same form, using coefficients whose values have yet to be determined.**

$$y_p = A_n x^n + A_{n-1} x^{n-1} + A_{n-2} x^{n-2} + \ldots + A_1 x + A_0$$

2. **If $g(x)$ is the sum of terms, $g_1(x)$, $g_2(x)$, $g_3(x)$ and so on, then you can break the problem into various subproblems.**

$$ay'' + by' + cy = g_1(x)$$
$$ay'' + by' + cy = g_2(x)$$
$$ay'' + by' + cy = g_3(x)$$

The particular solution is the sum of the solutions of these subproblems.

3. Substitute $y_p$ into the differential equation and solve for the undetermined coefficients.

4. Find the general solution, $y_h = c_1 y_1 + c_2 y_2$, of the associated homogeneous differential equation.

5. The general solution of the nonhomogeneous differential equation is the sum of $y_h$ and $y_p$.

6. Use the initial conditions to solve for $c_1$ and $c_2$, if the problem calls for that.

Take a look at the following example. Then, if you're game, try the following practice problems that ask you to solve for the general solution when $g(x)$ is a polynomial.

**Q.** Find the general solution to this differential equation:

$$y'' = 12x^2 + 12x - 2$$

where

$$y(0) = 1$$

and

$$y'(0) = 3$$

**A.** $y = x^4 + 2x^3 - x^2 + 3x + 1$

1. The homogeneous equation is simply

$$y'' = 0$$

2. You can get the solution by integrating

$$y' = c_1$$

and then integrating again to get $y_h$

$$y_h = c_1 x + c_2$$

So the solution to the homogeneous equation is $y_h = c_1 x + c_2$.

3. Now you need to find the solution to the nonhomogeneous equation. The $g(x)$ term is $12x^2 + 12x - 1$, so you can assume that the particular solution has a similar form:

$$y_p = Ax^2 + Bx + C$$

where $A$, $B$, and $C$ are constant coefficients that you must determine.

4. Uh oh. Your assumed form of $y_p$ has terms in common with $y_h$, the general solution of the homogeneous equation:

$$y_h = c_1 x + c_2$$
$$y_p = Ax^2 + Bx + C$$

Both of these equations have an $x$ term and a constant term. When $y_h$ and $y_p$ have terms in common — differing only by a multiplicative constant — that's not good because those terms are really part of the same solution.

5. Handle this issue by multiplying $y_p$ by successive powers of $x$ until you don't have any terms of the same power as in $y_h$. For example, multiply $y_p$ by $x$ to get

$$y_p = Ax^3 + Bx^2 + Cx$$

6. You still have a problem, however, because the $Cx$ term overlaps with the $c_1 x$ term in $y_h$. Go ahead and multiply by $x$ again:

$$y_p = Ax^4 + Bx^3 + Cx^2$$

This result has no terms in common with the homogeneous solution, $y_h$, so now you're good to go on.

7. Substitute the assumed solution into the differential equation:

$$12Ax^2 + 6Bx + 2C = 12x^2 + 12x - 2$$

8. Then compare coefficients of like terms:

$$12A = 12$$

$$6B = 12$$

$$2C = -2$$

So

$$A = 1$$

$$B = 2$$

$$C = -1$$

9. Here's your particular solution:

$$y_p = x^4 + 2x^3 - 1x^2$$

10. Now you can determine that your general solution is

$$y = y_h + y_p = c_1 x + c_2 + x^4 + 2x^3 - x^2$$

or, rearranging, you get

$$y = y_h + y_p = x^4 + 2x^3 - x^2 + c_1 x + c_2$$

11. You can find $c_1$ and $c_2$ by using the initial conditions. Substituting $y(0) = 1$ gives you

$$y(0) = 1 = c_2$$

so $c_2 = 1$.

12. Take the derivative of the general solution:

$$y' = 4x^3 + 3x^2 - 2x + c_1$$

13. Then substitute the initial condition, $y'(0) = 3$, to get

$$y'(0) = 3 = c_1$$

14. The general solution is thus

$$y = x^4 + 2x^3 - x^2 + 3x + 1$$

---

**7.** Find the general solution to the following nonhomogeneous second order differential equation:

$$y'' + 3y' + 2y = 4x$$

*Solve It*

**8.** Solve for the general solution to this equation:

$$y'' + 4y' + 3y = 3x + 10$$

*Solve It*

**9.** Find the general solution to the following nonhomogeneous second order differential equation:

$$y'' + 5y' + 6y = 12x - 2$$

*Solve It*

**10.** Solve for the general solution to this equation:

$$y'' + 6y' + 5y = 5x + 16$$

*Solve It*

# Solving Equations with a Nonhomogeneous Term That Involves Sines and Cosines

The third form $g(x)$ can take on is a combination of sines and cosines. Take a look at this differential equation:

$$y'' + 3y' + 2y = \sin(x)$$

Of course, the general solution is the sum of the homogeneous solution and a particular solution:

$$y = y_h + y_p$$

Because $g(x) = \sin(x)$ in this case, you can make an educated guess that

$$y_p = A \sin(x) + B \cos(x)$$

where $A$ and $B$ are undetermined coefficients. How do you find $A$ and $B$? Simply plug $y_p$ into the differential equation and then solve to get your coefficients.

The following problems give you practice using the method of undetermined coefficients to solve for the general solution to the nonhomogeneous equation when $g(x)$ includes sines and cosines.

**Q.** Find the general solution to this differential equation:

$$y'' + 3y' + 2y = 10 \sin (x)$$

where

$$y(0) = -1$$

and

$$y'(0) = -2$$

**A.** $y = e^{-x} + e^{-2x} + \sin (x) - 3 \cos (x)$

1. Start by getting the homogeneous version of the differential equation:

$$y'' + 3y' + 2y = 0$$

2. Assume that the solution to the homogeneous differential equation is of the form $y = e^{rx}$. When you substitute that into the differential equation, you get the following as the characteristic equation:

$$r^2 + 3r + 2 = 0$$

3. Factor the characteristic equation this way:

$$(r + 1)(r + 2) = 0$$

4. Determine that the roots, $r_1$ and $r_2$, of the characteristic equation are $-1$ and $-2$, giving you

$$y_1 = e^{-x} \text{ and } y_2 = e^{-2x}$$

5. Therefore, the solution to the homogeneous differential equation is given by

$$y_h = c_1 e^{-x} + c_2 e^{-2x}$$

6. Now you need a particular solution to the differential equation:

$$y'' + 3y' + 2y = 10 \sin (x)$$

7. Assume that the particular solution is of the form

$$y_p = A \sin (x) + B \cos (x)$$

8. Then plug the $A \sin (x)$ term into the left side of the differential equation to get

$$y'' + 3y' + 2y = -A \sin (x) + 3A \cos (x) + 2A \sin (x)$$

9. Now plug the $B \cos (x)$ term into the left side of the differential equation:

$$y'' + 3y' + 2y = -B \cos (x) - 3B \sin (x) + 2B \cos (x)$$

10. So you can write the differential equation as follows:

$$-A \sin (x) + 3A \cos (x) + 2A \sin (x) - B \cos (x) - 3B \sin (x) + 2B \cos (x) = 10 \sin (x)$$

which means that

$$3A \cos (x) - B \cos (x) + 2B \cos (x) = 0$$

and

$$-A \sin (x) + 2A \sin (x) - 3B \sin (x) = 10 \sin (x)$$

11. Dividing by $\sin (x)$ and $\cos (x)$ as appropriate gives you the first equation:

$$3A - B + 2B = 3A + B = 0$$

as well as the second:

$$-A + 2A - 3B = A - 3B = 10$$

12. Add three times the first equation to the second one:

$$9A + A = 10$$

Looks like $A = 1$.

13. Use the second equation to find that

$$A - 3B = 1 - 3B = 10$$

Guess what? $B = -3$.

14. So the particular solution is

$$y_p = \sin (x) - 3 \cos (x)$$

15. The general solution is

$$y = y_h + y_p$$

so that's

$$y = c_1 e^{-x} + c_2 e^{-2x} + \sin (x) - 3 \cos (x)$$

16. Use the initial conditions to find the first equation:

$$y(0) = -1 = c_1 + c_2 + \sin(0) - 3\cos(0) = c_1 + c_2 - 3$$

(so you get $c_1 + c_2 = 2$)

and the second equation:

$$y'(0) = -2 = -c_1 - 2c_2 + \cos(0) = -c_1 - 2c_2 + 1$$

(which gives you $-c_1 - 2c_2 = -3$)

17. Add the first equation to the second equation to get

$$-c_2 = -1$$

Tada! $c_2 = 1$

18. Plug that result into the first equation to get

$$1 + c_2 = 2$$

so $c_1 = 1$.

19. All of that means the general solution is

$$y = e^{-x} + e^{-2x} + \sin(x) - 3\cos(x)$$

---

**11.** Find the solution to the following non-homogeneous second order differential equation:

$$y'' + 5y' + 6y = 10\sin(x)$$

where

$$y(0) = 1$$

and

$$y'(0) = -4$$

*Solve It*

**12.** Solve for the general solution of this equation:

$$y'' + 4y' + 3y = 5\cos(x)$$

where

$$y(0) = \tfrac{1}{2}$$

and

$$y'(0) = -6$$

*Solve It*

# Answers to Nonhomogeneous Linear Second Order Differential Equation Problems

Following are the answers to the practice questions presented throughout this chapter. Each one is worked out step by step so that if you messed one up along the way, you can more easily see where you took a wrong turn.

**1** **What's the general solution to this nonhomogeneous second order differential equation?**

$$y'' + 3y' + 2y = 6e^x$$

**Solution:** $y = c_1 e^{-x} + c_2 e^{-2x} + e^x$

1. First, find the homogeneous version of the original equation:

   $$y'' + 3y' + 2y = 0$$

2. Assume that the solution to the homogeneous differential equation is of the form $y = e^{rx}$. When you substitute that solution into the equation, you get the characteristic equation

   $$r^2 + 3r + 2 = 0$$

3. Go ahead and factor that out as follows:

   $$(r + 1)(r + 2) = 0$$

4. If you determine that the roots, $r_1$ and $r_2$, of the characteristic equation are –1 and –2, you know that

   $$y_1 = e^{-x} \text{ and } y_2 = e^{-2x}$$

5. Thus, the solution to the homogeneous differential equation is given by

   $$y = c_1 e^{-x} + c_2 e^{-2x}$$

6. Now you need a particular solution to the differential equation:

   $$y'' + 3y' + 2y = 6e^x$$

   Note that $g(x)$ has the form $e^x$ here, so assume that the particular solution has the form

   $$y_p(x) = Ae^x$$

7. Substitute $y_p(x)$ into the equation:

   $$Ae^x + 3Ae^x + 2Ae^x = 6e^x$$

8. Cancel out the $e^x$ term:

   $$A + 3A + 2A = 6$$

   or $6A = 6$, so $A = 1$.

9. Your particular solution is

   $$y_p(x) = e^x$$

10. Because the general solution of the nonhomogeneous equation that you started with is the sum of the corresponding homogeneous equation's general solution and a particular solution of the nonhomogeneous equation, you should get the following as the solution:

    $$y = y_h + y_p$$

    which is actually

    $$y = c_1 e^{-x} + c_2 e^{-2x} + e^x$$

> 2 **Find the general solution to the following nonhomogeneous second order differential equation:**
>
> $$y'' + 4y' + 3y = 30e^{2x}$$
>
> **Solution:** $y = c_1e^{-x} + c_2e^{-3x} + 2e^{2x}$
>
> 1. Start by getting the homogeneous version of the differential equation:
>
>    $$y'' + 4y' + 3y = 0$$
>
> 2. Go ahead and assume that the solution to this equation is of the form $y = e^{rx}$ and that when you substitute this solution into the equation you get the following as your characteristic equation:
>
>    $$r^2 + 4r + 3 = 0$$
>
> 3. Factor the characteristic equation this way:
>
>    $$(r + 1)(r + 3) = 0$$
>
> 4. Determining that the roots, $r_1$ and $r_2$, of the characteristic equation are $-1$ and $-3$ gives you
>
>    $$y_1 = e^{-x} \text{ and } y_2 = e^{-3x}$$
>
> 5. Therefore, you know that the solution to the homogeneous differential equation is given by
>
>    $$y = c_1e^{-x} + c_2e^{-3x}$$
>
> 6. Now you need a particular solution to the equation:
>
>    $$y'' + 4y' + 3y = 30e^{2x}$$
>
>    Note that $g(x)$ has the form $e^{2x}$ here, so assume that the particular solution has this form:
>
>    $$y_p(x) = Ae^{2x}$$
>
> 7. Substituting $y_p(x)$ into the differential equation gives you
>
>    $$4Ae^{2x} + 8Ae^{2x} + 3Ae^{2x} = 30e^{2x}$$
>
> 8. If you cancel out the $e^{2x}$ term, you get
>
>    $$4A + 8A + 3A = 30$$
>
>    which is
>
>    $$15A = 30$$
>
>    and that means $A = 2$, which gives you your particular solution of
>
>    $$y_p(x) = 2e^{2x}$$
>
> 9. Put it all together and you get this equation as the answer:
>
>    $$y = y_h + y_p$$
>
>    or
>
>    $$y = c_1e^{-x} + c_2e^{-3x} + 2e^{2x}$$

> 3 **Solve for the general solution to this equation:**
>
> $$y'' + 5y' + 6y = 36e^{x}$$
>
> **Solution:** $y = c_1e^{-2x} + c_2e^{-3x} + 3e^{x}$
>
> 1. First, find the homogeneous version of the original equation:
>
>    $$y'' + 5y' + 6y = 0$$

2. Assume that the solution to the homogeneous differential equation is of the form $y = e^{rx}$. When you substitute that solution into the equation, you get the characteristic equation

$$r^2 + 5r + 6 = 0$$

3. Go ahead and factor that out as follows:

$$(r + 2)(r + 3) = 0$$

4. If you determine that the roots, $r_1$ and $r_2$, of the characteristic equation are $-2$ and $-3$, you know that

$$y_1 = e^{-2x} \text{ and } y_2 = e^{-3x}$$

5. Thus, the solution to the homogeneous differential equation is given by

$$y = c_1 e^{-2x} + c_2 e^{-3x}$$

6. Now you need a particular solution to the differential equation:

$$y'' + 5y' + 6y = 36e^x$$

Note that $g(x)$ has the form $e^x$ here, so assume that the particular solution has the form

$$y_p(x) = Ae^x$$

7. Substitute $y_p(x)$ into the equation:

$$Ae^x + 5Ae^x + 6Ae^x = 36e^x$$

8. Cancel out the $e^x$ term:

$$A + 5A + 6A = 36$$

or $12A = 36$, so $A = 3$.

9. Your particular solution is

$$y_p(x) = 3e^x$$

10. When you put all that together, you should get the following as the solution:

$$y = y_h + y_p$$

which is actually

$$y = c_1 e^{-2x} + c_2 e^{-3x} + 3e^x$$

**4** **What's the general solution to this nonhomogeneous second order differential equation?**

$$y'' + 2y' + y = 8e^x$$

**Solution:** $y = c_1 e^{-x} + c_2 x e^{-x} + 2e^x$

1. Start by getting the homogeneous version of the differential equation:

$$y'' + 2y' + y = 0$$

2. Go ahead and assume that the solution to this equation is of the form $y = e^{rx}$ and that when you substitute this solution into the equation you get the following as your characteristic equation:

$$r^2 + 2r + 1 = 0$$

3. Factor the characteristic equation this way:

$$(r + 1)(r + 1) = 0$$

4. Determining that the roots, $r_1$ and $r_2$, of the characteristic equation are $-1$ and $-1$ (which means you have real, identical roots) gives you

$$y_1 = e^{-x} \text{ and } y_2 = xe^{-x}$$

5. Therefore, you know that the solution to the homogeneous differential equation is given by

$$y = c_1 e^{-x} + c_2 x e^{-x}$$

6. Now you need a particular solution to the equation:

$$y'' + 2y' + y = 8e^x$$

Note that $g(x)$ has the form $e^x$ here, so assume that the particular solution has this form:

$$y_p(x) = Ae^x$$

7. Substituting $y_p(x)$ into the differential equation gives you

$$Ae^x + 2Ae^x + Ae^x = 8e^x$$

8. If you cancel out the $e^x$ term, you get

$$A + 2A + A = 8$$

which is

$$4A = 8$$

and that means $A = 2$, which gives you your particular solution of

$$y_p(x) = 2e^x$$

9. Put it all together and you get this equation as the answer:

$$y = y_h + y_p$$

or

$$y = c_1 e^{-x} + c_2 x e^{-x} + 2e^x$$

**5** **Find the general solution to the following nonhomogeneous second order differential equation:**

$$y'' + 6y' + 8y = 70e^{3x}$$

**Solution: $y = c_1 e^{-2x} + c_2 e^{-4x} + 2e^{3x}$**

1. First, find the homogeneous version of the original equation:

$$y'' + 6y' + 8y = 0$$

2. Assume that the solution to the homogeneous differential equation is of the form $y = e^{rx}$. When you substitute that solution into the equation, you get the characteristic equation

$$r^2 + 6r + 8 = 0$$

3. Go ahead and factor that out as follows:

$$(r + 2)(r + 4) = 0$$

4. If you determine that the roots, $r_1$ and $r_2$, of the characteristic equation are $-2$ and $-4$, you know th

$$y_1 = e^{-2x} \text{ and } y_2 = e^{-4x}$$

5. Thus, the solution to the homogeneous differential equation is given by

$$y = c_1 e^{-2x} + c_2 e^{-4x}$$

6. Now you need a particular solution to the differential equation:

$$y'' + 6y' + 8y = 70e^{3x}$$

Note that $g(x)$ has the form $e^x$ here, so assume that the particular solution has the form

$$y_p(x) = Ae^{3x}$$

7. Substitute $y_p(x)$ into the equation:

$$9Ae^{3x} + 18Ae^{3x} + 8Ae^{3x} = 70e^{3x}$$

8. Cancel out the $e^x$ term:

$$9A + 18A + 8A = 70$$

or $35A = 70$, so $A = 2$.

9. Your particular solution is

$$y_p(x) = 2e^{3x}$$

10. When you put all that together, you should get the following as the solution:

$$y = y_h + y_p$$

which is actually

$$y = c_1e^{-2x} + c_2e^{-4x} + 2e^{3x}$$

**6** **Solve for the general solution to this equation:**

$$y'' + 6y' + 5y = 36e^x$$

**Solution:** $y = c_1e^{-x} + c_2e^{-5x} + 3e^x$

1. Start by getting the homogeneous version of the differential equation:

$$y'' + 6y' + 5y = 0$$

2. Go ahead and assume that the solution to this equation is of the form $y = e^{rx}$ and that when you substitute this solution into the equation you get the following as your characteristic equation:

$$r^2 + 6r + 5 = 0$$

3. Factor the characteristic equation this way:

$$(r + 1)(r + 5) = 0$$

4. If you determine that the roots, $r_1$ and $r_2$, of the characteristic equation are $-1$ and $-5$, you know that

$$y_1 = e^{-x} \text{ and } y_2 = e^{-5x}$$

5. Therefore, you know that the solution to the homogeneous differential equation is given by

$$y = c_1e^{-x} + c_2e^{-5x}$$

6. Now you need a particular solution to the equation:

$$y'' + 6y' + 5y = 36e^x$$

Note that $g(x)$ has the form $e^x$ here, so assume that the particular solution has this form:

$$y_p(x) = Ae^x$$

7. Substituting $y_p(x)$ into the differential equation gives you

$$Ae^x + 6Ae^x + 5Ae^x = 36e^x$$

8. If you cancel out the $e^x$ term, you get

$$A + 6A + 5A = 36$$

which is

$$12A = 36$$

and that means $A = 2$, which gives you your particular solution of

$$y_p(x) = 3e^x$$

9. Put it all together and you get this equation as the answer:

$$y = y_h + y_p$$

or

$$y = c_1 e^{-x} + c_2 e^{-5x} + 3e^x$$

**7** **Find the general solution to the following nonhomogeneous second order differential equation:**

$$y'' + 3y' + 2y = 4x$$

**Solution:** $y = c_1 e^{-x} + c_2 e^{-2x} + 2x - 3$

1. Get the homogeneous version of the equation first:

$$y'' + 3y' + 2y = 0$$

2. If you assume that the solution to the homogeneous equation is of the form $y = e^{rx}$, you get the following characteristic equation when you substitute that solution in:

$$r^2 + 3r + 2 = 0$$

3. Factor out the characteristic equation as follows:

$$(r + 1)(r + 2) = 0$$

4. Then determine that the roots, $r_1$ and $r_2$, of the characteristic equation are $-1$ and $-2$. Doing so gives you

$$y_1 = e^{-x} \text{ and } y_2 = e^{-2x}$$

5. So the solution to the homogeneous equation is given by

$$y = c_1 e^{-x} + c_2 e^{-2x}$$

6. Hold up! You're not done yet. You still need a particular solution to the differential equation:

$$y'' + 3y' + 2y = 4x$$

Note that here $g(x)$ has the form of a polynomial, so you can assume that the particular solution has the form

$$y_p(x) = Ax^2 + Bx + C$$

7. Great. Now substitute $y_p(x)$ into the differential equation:

$$2A + 6Ax + 3B + 2Ax^2 + 2Bx + 2C = 4x$$

8. There's no term in $x^2$ on the right, so $A = 0$, giving you

$$3B + 2Bx + 2C = 4x$$

9. Looking at the coefficient of $x$ gives you

$$2B = 4, \text{ so } B = 2.$$

10. Now you can take a look at the remaining constant terms; doing so gives you

$$3B + 2C = 0$$

Because $B = 2$, that means $C = -3$.

11. So the particular solution is

$$y_p(x) = 2x - 3$$

12. Of course, you haven't forgotten that the general solution of the nonhomogeneous equation you started with is the sum of the corresponding homogeneous equation's general solution and a particular solution of the nonhomogeneous equation. That means you've found this solution:

$$y = y_h + y_p$$

which can also be written as

$$y = c_1 e^{-x} + c_2 e^{-2x} + 2x - 3$$

*S* **Solve for the general solution to this equation:**

$$y'' + 4y' + 3y = 3x + 10$$

**Solution: $y = c_1 e^{-x} + c_2 e^{-3x} + x + 2$**

1. Start by finding the homogeneous version of the original differential equation:

$$y'' + 4y' + 3y = 0$$

2. Assuming that the solution to the homogeneous equation is of the form $y = e^{rx}$ means that when you substitute that solution into the differential equation, you get this characteristic equation:

$$r^2 + 4r + 3 = 0$$

3. Here's how to factor it:

$$(r + 1)(r + 3) = 0$$

4. Now go ahead and determine that the roots, $r_1$ and $r_2$, of the characteristic equation are $-1$ and $-3$, which gives you

$$y_1 = e^{-x} \text{ and } y_2 = e^{-3x}$$

5. Thus, the solution to the homogeneous differential equation is given by

$$y = c_1 e^{-x} + c_2 e^{-3x}$$

6. Now you need a particular solution to this equation:

$$y'' + 4y' + 3y = 3x + 10$$

Note that $g(x)$ has the form of a polynomial here, which means you can safely bet that the particular solution has this form:

$$y_p(x) = Ax^2 + Bx + C$$

7. Substituting $y_p(x)$ into the equation gives you

$$2A + 8Ax + 4B + 3Ax^2 + 3Bx + 3C = 3x + 10$$

8. Without an $x^2$ term on the right, $A = 0$, giving you

$$4B + 3Bx + 3C = 3x + 10$$

9. Look at the coefficient of $x$ to get

$$3B = 3, \text{ which means } B = 1.$$

10. A quick look at the remaining constant terms gives you

$$4B + 3C = 10$$

What do you know? $B = 1$, so $C = 2$.

11. Your particular solution is

$$y_p(x) = x + 2$$

12. Put it all together to get the following equation as your solution:

$$y = y_h + y_p$$

so

$$y = c_1e^{-x} + c_2e^{-3x} + x + 2$$

**9** **Find the general solution to the following nonhomogeneous second order differential equation:**

$$y'' + 5y' + 6y = 12x - 2$$

**Solution:** $y = c_1e^{-2x} + c_2e^{-3x} + 2x - 2$

1. Get the homogeneous version of the equation first:

$$y'' + 5y' + 6y = 0$$

2. If you assume that the solution to the homogeneous equation is of the form $y = e^{rx}$, you get the following characteristic equation when you substitute that solution in:

$$r^2 + 5r + 6 = 0$$

3. Factor out the characteristic equation as follows:

$$(r + 2)(r + 3) = 0$$

4. Then determine that the roots, $r_1$ and $r_2$, of the characteristic equation are $-2$ and $-3$. Doing so gives you

$$y_1 = e^{-2x} \text{ and } y_2 = e^{-3x}$$

5. So the solution to the homogeneous equation is given by

$$y = c_1e^{-2x} + c_2e^{-3x}$$

6. Hold up! You're not done yet. You still need a particular solution to the differential equation:

$$y'' + 5y' + 6y = 12x - 2$$

Note that here $g(x)$ has the form of a polynomial, so you can assume that the particular solution has the form

$$y_p(x) = Ax^2 + Bx + C$$

7. Great. Now substitute $y_p(x)$ into the differential equation:

$$2A + 10Ax + 5B + 6Ax^2 + 6Bx + 6C = 12x - 2$$

8. There's no term in $x^2$ on the right, so $A = 0$, giving you

$$5B + 6Bx + 6C = 12x - 2$$

9. Looking at the coefficient of $x$ gives you

$$6B = 12, \text{ so } B = 2.$$

10. Now you can take a look at the remaining constant terms; doing so gives you

$$5B + 6C = -2$$

Because $B = 2$, that means $C = -2$.

11. So the particular solution is

$y_p(x) = 2x - 2$

12. When you put that all together, you should get this as the general solution:

$y = y_h + y_p$

which can also be written as

$y = c_1 e^{-2x} + c_2 e^{-3x} + 2x - 2$

**10** **Solve for the general solution to this equation:**

$y'' + 6y' + 5y = 5x + 16$

**Solution: $y = c_1 e^{-x} + c_2 e^{-5x} + x + 2$**

1. Start by finding the homogeneous version of the original differential equation:

$y'' + 6y' + 5y = 0$

2. Assuming that the solution to the homogeneous equation is of the form $y = e^{rx}$ means that when you substitute that solution into the equation, you get this characteristic equation:

$r^2 + 6r + 5 = 0$

3. Here's how to factor it:

$(r + 1)(r + 5) = 0$

4. Now go ahead and determine that the roots, $r_1$ and $r_2$, of the characteristic equation are $-1$ and $-5$, which gives you

$y_1 = e^{-x}$ and $y_2 = e^{-5x}$

5. Thus, the solution to the homogeneous differential equation is given by

$y = c_1 e^{-x} + c_2 e^{-5x}$

6. Now you need a particular solution to this equation:

$y'' + 6y' + 5y = 5x + 16$

Note that $g(x)$ has the form of a polynomial here, which means you can safely bet that the particular solution has this form:

$y_p(x) = Ax^2 + Bx + C$

7. Substituting $y_p(x)$ into the equation gives you

$2A + 12Ax + 6B + 5Ax^2 + 5Bx + 5C = 5x + 16$

8. Without an $x^2$ term on the right, $A = 0$, giving you

$6B + 5Bx + 5C = 5x + 16$

9. Look at the coefficient of $x$ to get

$5B = 5$, which means $B = 1$.

10. A quick look at the remaining constant terms gives you

$6B + 5C = 16$

What do you know? $B = 1$, so $C = 2$.

11. Your particular solution is

$$y_p(x) = x + 2$$

12. Put it all together to get the following equation as your solution:

$$y = y_h + y_p$$

so

$$y = c_1 e^{-x} + c_2 e^{-5x} + x + 2$$

**11** **Find the solution to the following nonhomogeneous second order differential equation:**

$$y'' + 5y' + 6y = 10 \sin (x)$$

**where**

$$y(0) = 1$$

**and**

$$y'(0) = -4$$

**Solution:** $y = e^{-2x} + e^{-3x} + \sin (x) - \cos (x)$

1. First, find the homogeneous version of the differential equation:

$$y'' + 5y' + 6y = 0$$

2. Work under the assumption that the solution to the homogeneous equation is of the form $y = e^{rx}$. Doing so means that when you substitute that solution into the differential equation, you get this characteristic equation:

$$r^2 + 5r + 6 = 0$$

3. The next step is to factor the characteristic equation as follows:

$$(r + 2)(r + 3) = 0$$

4. Then determine that the roots, $r_1$ and $r_2$, of the characteristic equation are $-2$ and $-3$, which gives you

$$y_1 = e^{-2x} \text{ and } y_2 = e^{-3x}$$

5. You now know that the solution to the homogeneous equation is given by

$$y_h = c_1 e^{-2x} + c_2 e^{-3x}$$

6. Next, you need a particular solution to the differential equation:

$$y'' + 5y' + 6y = 10 \sin (x)$$

7. Assume the particular solution is of this form:

$$y_p = A \sin (x) + B \cos (x)$$

8. Plug the $A \sin (x)$ term into the left side of the equation:

$$y'' + 5y' + 6y = -A \sin (x) + 5A \cos (x) + 6A \sin (x)$$

9. Then plug the $B \cos (x)$ term into the left side of the equation:

$$y'' + 5y' + 6y = -B \cos (x) - 5B \sin (x) + 6B \cos (x)$$

10. Hmmm. It appears you can write the differential equation this way:

$$-A \sin (x) + 5A \cos (x) + 6A \sin (x) - B \cos (x) - 5B \sin (x) + 6B \cos (x) = \sin (x)$$

which means that

$$5A \cos (x) - B \cos (x) + 6B \cos (x) = 0$$

and

$$-A \sin (x) + 6A \sin (x) - 5B \sin (x) = 10 \sin (x)$$

11. Dividing by $\sin (x)$ and $\cos (x)$ as appropriate gives you the first equation:

$$5A - B + 6B = 5A + 5B = 0$$

as well as the second one:

$$-A + 6A - 5B = 5A - 5B = 10$$

12. Add the first equation to the second:

$$10A = 10, \text{ so } A = 1.$$

13. Use the second equation to find the following:

$$5A - 5B = 5 - 5B = 10$$

so $B = -1$.

14. Therefore, the particular solution is

$$y_p = \sin (x) - \cos (x)$$

and the general solution is

$$y = y_h + y_p$$

so that's

$$y = c_1 e^{-2x} + c_2 e^{-3x} + \sin (x) - \cos (x)$$

15. Use the initial conditions to find the first equation:

$$y(0) = 1 = c_1 e^{-x} + c_2 e^{-2x} + \sin (x) - \cos (x) = c_1 + c_2 - 1$$

and the second equation:

$$y'(0) = -4 = -2c_1 - 3c_2 + 1$$

16. Add twice the first equation to the second one to get

$$-c_2 + 1 = 0$$

so $c_2 = 1$.

17. Plug that result into the first equation to get

$$-2c_2 - 3 = -5$$

so $c_1 = 1$.

18. That means the general solution is

$$y = e^{-2x} + e^{-3x} + \sin (x) - \cos (x)$$

12 **Solve for the general solution of this equation:**

$$y'' + 4y' + 3y = 5 \cos (x)$$

**where**

$$y(0) = \text{ }^{7}/_{2}$$

**and**

$$y'(0) = -6$$

**Solution: $y = e^{-x} + 2e^{-3x} + \sin (x) + \cos (x)/2$**

1. Start by getting the homogeneous version of the original differential equation:

$$y'' + 4y' + 3y = 0$$

2. Go ahead and assume that the solution to the homogeneous equation is of the form $y = e^{rx}$. When you substitute that into the equation, you get this characteristic equation:

$$r^2 + 4r + 3 = 0$$

3. Factor that out:

$$(r + 1)(r + 3) = 0$$

4. Determine that the roots, $r_1$ and $r_2$, of the characteristic equation are –1 and –3. Doing so gives you

$$y_1 = e^{-x} \text{ and } y_2 = e^{-3x}$$

5. So the solution to the homogeneous differential equation is given by

$$y_h = c_1 e^{-x} + c_2 e^{-3x}$$

6. That's all well and good, but you still need a particular solution to the differential equation

$$y'' + 4y' + 3y = 5 \cos (x)$$

7. Assume that the particular solution is of this form:

$$y_p = A \sin (x) + B \cos (x)$$

8. Then plug $A \sin (x)$ into the left side of the differential equation to get

$$y'' + 4y' + 3y = -A \sin (x) + 4A \cos (x) + 3A \sin (x)$$

9. Next, plug $B \cos (x)$ into the left side of the equation to get

$$y'' + 4y' + 3y = -B \cos (x) - 4B \sin (x) + 3B \cos (x)$$

10. Surprise! Turns out you can write the differential equation this way:

$$-A \sin (x) + 4A \cos (x) + 3A \sin (x) - B \cos (x) - 4B \sin (x) + 3B \cos (x) = \cos (x)$$

which means that

$$4A \cos (x) - B \cos (x) + 3B \cos (x) = 5 \cos (x)$$

and

$$-A \sin (x) + 3A \sin (x) - 4B \sin (x) = 0$$

11. Dividing by $\sin (x)$ and $\cos (x)$ as appropriate gives you this as the first equation:

$$4A - B + 3B = 4A + 2B = 5$$

and this as the second equation:

$$-A + 3A - 4B = 2A - 4B = 0$$

12. Add twice the first equation to the second one; the result is

$10A = 10$, so $A = 1$.

13. Use the second equation to find that

$2A - 4B = 0$, so $B = \frac{1}{2}$

which means the particular solution is

$y_p = \sin(x) + \cos(x)/2$

and the general solution is

$y = y_h + y_p$

which is actually

$y = c_1 e^{-x} + c_2 e^{-3x} + \sin(x) + \cos(x)/2$

14. Use the initial conditions to find the first equation:

$y(0) = \frac{7}{2} = c_1 e^{-x} + c_2 e^{-3x} + \sin(x) + \cos(x)/2 = c_1 + c_2 + \frac{1}{2}$

and the second equation:

$y'(0) = -6 = -c_1 - 3c_2 + 1$.

15. Rewrite the first equation as follows:

$3 = c_1 + c_2$

and the second equation like this:

$-7 = -c_1 - 3c_2$

16. Add the first equation to the second one to get

$-4 = -2c_2$, which means that $c_2 = 2$.

17. Plug that result into the first equation to get

$-7 = -c_1 - 6$, so $c_1 = 1$.

18. Consequently, your general solution is

$y = e^{-x} + 2e^{-3x} + \sin(x) + \cos(x)/2$

# Chapter 6

# Handling Homogeneous Linear Higher Order Differential Equations

## In This Chapter

▶ Reviewing the higher order process with real and distinct roots

▶ Adding complexity with complex roots

▶ Avoiding double duty with duplicate roots

*T*his chapter is where you get to practice tackling higher order differential equations, where $n > 2$. (***Note:*** Higher order equations are sometimes referred to as $n$th order equations.) A general linear higher order differential equation looks like this:

$$\frac{d^n y}{dx^n} + p_1(x)\frac{d^{n-1}y}{dx^{n-1}} + p_2(x)\frac{d^{n-2}y}{dx^{n-2}} + \dots + p_{n-1}(x)\frac{dy}{dx} + p_n(x)y = g(x)$$

Solving higher order differential equations where $n = 3$ or more is a lot like solving differential equations of the first or second order, with two exceptions: You need more integrations, and you have to solve larger systems of simultaneous equations to meet the initial conditions.

Every linear higher order differential equation you encounter in this chapter has constant coefficients, and the main way you can plan to tackle these problems is by attempting a solution of the form

$y = e^{rx}$

Substituting in this attempted solution results in a characteristic equation in powers of $r$, just as it does for the linear second order differential equations covered in Chapters 4 and 5. The problem here is that you're dealing with cubic (or higher!) characteristic equations, as well as $3 \times 3$ systems of simultaneous equations to handle the initial conditions.

Whenever you're facing a characteristic equation that's tough to solve by hand, I recommend turning to a Web-based equation solver. You can find a good one at www.quickmath.com. From the home page, look for the Equations listing on the left-hand navigation bar. Then click the Solve link under the Equations listing. You can also solve systems of simultaneous equations online at math.cowpi.com/systemsolver. To make your life a little easier, you may want to refer to these Web sites as you solve the practice problems in this chapter.

As with linear second order differential equations, the characteristic equation you find can have three types of roots:

- ✔ Real and distinct roots
- ✔ Complex roots
- ✔ Real and identical roots

In this chapter, you get to try your hand at each of these possibilities in homogeneous linear higher order differential equations.

# Distinctly Different: Working with Real and Distinct Roots

In this section, you practice the case where the characteristic equation has real and distinct roots first — that is, the roots aren't imaginary, and they're not the same.

Here's a linear second order differential equation that's homogeneous and has constant coefficients:

$$y'' + 3y' + 2 = 0$$

Given this equation's form, you can safely bet that the solutions are something like

$$y = e^{rx}$$

Plugging that solution into the differential equation gives you

$$r^2 e^{rx} + 3re^{rx} + 2e^{rx} = 0$$

and dividing by $e^{rx}$ gives you

$$r^2 + 3r + 2 = 0$$

Surprise! There's your characteristic equation, which you can solve with the quadratic equation to get

$$(r + 1)(r + 2) = 0$$

So the characteristic equation's roots are –1 and –2, giving you these two solutions:

$$y = e^{-x} \text{ and } y = e^{-2x}$$

The process is similar for higher order differential equations, but the algebra is a little tougher because the characteristic equation is of a higher order. See what I mean in the following example problem and then try to solve a few of the practice problems on your own.

**Q.** Find the solution to this differential equation:

$$y''' - 6y'' + 11y' - 6y = 0$$

with these initial conditions:

$$y(0) = 9$$

$$y'(0) = 20$$

$$y''(0) = 50$$

**A.** $y = 2e^x + 3e^{2x} + 4e^{3x}$

1. This differential equation has constant coefficients, so you can start by assuming a solution of the form

$$y = e^{rx}$$

2. Plugging your attempted solution into the differential equation gives you

$$r^3 e^{rx} - 6r^2 e^{rx} + 11re^{rx} - 6e^{rx} = 0$$

3. Canceling out $e^{rx}$ leaves you with

$$r^3 - 6r^2 + 11r - 6 = 0$$

4. Now you have a cubic equation. Curb that racing pulse and take a minute to really look at the equation. See how you can factor it into the following?

$$(r - 1)(r - 2)(r - 3) = 0$$

If you're not a fan of factoring the characteristic equation by hand, try using the equation-solving function at www. quickmath.com. (Flip back to the chapter introduction to see exactly how to access that part of the Web site.)

5. The roots are

$$r_1 = 1, r_2 = 2, \text{ and } r_3 = 3$$

6. Because the roots are real and distinct, the solutions are as follows:

$$y_1 = e^x, y_2 = e^{2x}, \text{ and } y_3 = e^{3x}$$

7. Therefore, the general solution is

$$y = c_1 e^x + c_2 e^{2x} + c_3 e^{3x}$$

8. Now you can apply the initial conditions (about time, huh?). In addition to the form for $y$, you also need $y'$, which you can calculate as

$$y' = c_1 e^x + 2c_2 e^{2x} + 3c_3 e^{3x}$$

as well as $y''$, which you can calculate as

$$y'' = c_1 e^x + 4c_2 e^{2x} + 9c_3 e^{3x}$$

9. From the initial conditions, here are your three simultaneous equations in $c_1$, $c_2$, and $c_3$ that you must solve to find those coefficients:

$$y(0) = c_1 + c_2 + c_3 = 9$$

$$y'(0) = c_1 + 2c_2 + 3c_3 = 20$$

$$y''(0) = c_1 + 4c_2 + 9c_3 = 50$$

10. If you solve this system of three equations by hand, you should get

$$c_1 = 2, c_2 = 3, \text{ and } c_3 = 4$$

Of course, you can also take advantage of math.cowpi.com/systemsolver. Just click the $3 \times 3$ link and input the equations to solve the system.

11. So the solution of the differential equation with the initial conditions applied is

$$y = 2e^x + 3e^{2x} + 4e^{3x}$$

**1.** Find the solution to the following differential equation:

$$y''' + 7y'' + 14y' + 8y = 0$$

with these initial conditions:

$$y(0) = 3$$

$$y'(0) = -7$$

$$y''(0) = 14$$

*Solve It*

**2.** Solve this equation:

$$y''' + 8y'' + 19y' + 12y = 0$$

where

$$y(0) = 4$$

$$y'(0) = -12$$

$$y''(0) = 42$$

*Solve It*

**3.** Obtain the solution to this equation:

$$y''' + 9y'' + 26y' + 24y = 0$$

by using these initial conditions:

$$y(0) = 5$$

$$y'(0) = -16$$

$$y''(0) = 54$$

*Solve It*

**4.** Find the solution to the following differential equation:

$$y''' + 10y'' + 31y' + 30y = 0$$

with these initial conditions:

$$y(0) = 6$$

$$y'(0) = -19$$

$$y''(0) = 71$$

*Solve It*

**5.** Solve this equation:

$$y''' + 11y'' + 36y' + 36y = 0$$

where

$$y(0) = 5$$

$$y'(0) = -17$$

$$y''(0) = 67$$

Solve It

**6.** Obtain the solution to this equation:

$$y''' + 10y'' + 29y' + 20y = 0$$

by using these initial conditions:

$$y(0) = 7$$

$$y'(0) = -30$$

$$y''(0) = 142$$

Solve It

# A Cause for Complexity: Handling Complex Roots

What if a differential equation's characteristic equation has roots that are complex (meaning they involve the imaginary number $i$), such as

$$r_1 = i \text{ and } r_2 = -i$$

You can handle such a case with these two relationships:

$$e^{(\alpha + i\beta)x} = e^{\alpha x}(\cos \beta x + i \sin \beta x)$$

and

$$e^{(\alpha + i\beta)x} = e^{\alpha x}(\cos \beta x - i \sin \beta x)$$

Spend a few minutes reviewing the following example of how to solve this type of equation, or if you're feeling up to it, take advantage of the opportunity to work through some practice problems that feature linear higher order differential equations with complex roots.

**Q.** Find the solution to this differential equation:

$$y^{(4)} - y = 0$$

where

$$y(0) = 3$$

$$y'(0) = 1$$

$$y''(0) = -1$$

$$y'''(0) = -3$$

**A.** $y = e^{-x} + 2\cos x + 2\sin x$

1. You know that this differential equation has constant coefficients, so go ahead and assume a solution of the form

$$y = e^{rx}$$

2. Plug your solution into the differential equation to get

$$r^4 e^{rx} - e^{rx} = 0$$

3. Then cancel out $e^{rx}$:

$$r^4 - 1 = 0$$

4. You can factor the resulting characteristic equation into

$$(r^2 - 1)(r^2 + 1) = 0$$

5. So the roots of the characteristic equation are

$$r_1 = 1, r_2 = -1, r_3 = i, \text{ and } r_4 = -i$$

6. Use these relationships to make the solution easier to find:

$$e^{i\beta x} = \cos \beta x + i \sin \beta x$$

and

$$e^{-i\beta x} = \cos \beta x - i \sin \beta x$$

7. Okay, so you've determined that $y_3$ and $y_4$ can be expressed as a linear combination of sines and cosines (note that you can absorb the $i$ into a multiplicative constant). That means you have these solutions:

$$y_1 = e^x$$

$$y_2 = e^{-x}$$

$$y_3 = \cos x$$

$$y_4 = \sin x$$

8. So the general solution is

$$y = c_1 e^x + c_2 e^{-x} + c_3 \cos x + c_4 \sin x$$

9. To apply the initial conditions, you need to figure out $y'$:

$$y' = c_1 e^x - c_2 e^{-x} - c_3 \sin x + c_4 \cos x$$

and $y''$:

$$y'' = c_1 e^x + c_2 e^{-x} - c_3 \cos x - c_4 \sin x$$

as well as $y'''$:

$$y''' = c_1 e^x - c_2 e^{-x} + c_3 \sin x - c_4 \cos x$$

10. Substituting the forms for $y, y', y''$, and $y'''$ into the initial conditions gives you

$$y(0) = c_1 + c_2 + c_3 = 3$$

$$y'(0) = c_1 - c_2 + c_4 = 1$$

$$y''(0) = c_1 + c_2 - c_3 = -1$$

$$y'''(0) = c_1 - c_2 - c_4 = -3$$

11. Solve this $4 \times 4$ simultaneous equation system by hand or with an online tool such as the $4 \times 4$ system solver at `math.cowpi.com/systemsolver`:

$$c_1 = 0, c_2 = 1, c_3 = 2, \text{ and } c_4 = 2$$

12. After all that, the general solution with initial conditions applied is

$$y = e^{-x} + 2\cos x + 2\sin x$$

**7.** What's the solution to the following differential equation?

$$y^{(4)} - 16y = 0$$

where

$$y(0) = 3$$

$$y'(0) = 2$$

$$y''(0) = -4$$

$$y'''(0) = -24$$

*Solve It*

**8.** Find the solution to this differential equation:

$$y^{(4)} - 81y = 0$$

where

$$y(0) = 4$$

$$y'(0) = 12$$

$$y''(0) = -18$$

$$y'''(0) = -162$$

*Solve It*

# Identity Issues: Solving Equations When Identical Roots Are Involved

Identical roots are a no-brainer to spot, but they can be a bit messy to solve if you don't know what you're doing. Why? Well, if you have a differential equation whose characteristic equation has the roots –2, –2, –2, and –2, then those four –2s are an issue, because you can't just say the solutions are

$$y_1 = e^{-2x}$$
$$y_2 = e^{-2x}$$
$$y_3 = e^{-2x}$$
$$y_4 = e^{-2x}$$

Using all of these solutions would give you a general solution like

$$y = c_1 e^{-2x} + c_2 e^{-2x} + c_3 e^{-2x} + c_4 e^{-2x}$$

which is really equivalent to the following if you combine the constants:

$$y = ce^{-2x}$$

where $c = c_1 + c_2 + c_3 + c_4$.

You can handle such scenarios by adding powers of $x$. For example, if $y_1$ equals $c_1 e^{-2x}$, then you get

$$y_2 = c_2 x e^{-2x}$$
$$y_3 = c_3 x^2 e^{-2x}$$
$$y_4 = c_4 x^3 e^{-2x}$$

for the rest. So the general solution is

$$y = c_1 e^{-2x} + c_2 x e^{-2x} + c_3 x^2 e^{-2x} + c_4 x^3 e^{-2x}$$

Take a look at this example to see another linear higher order differential equation with identical roots being worked out and then try a few equations yourself.

***Q.*** Find the solution to this differential equation:

$$y^{(4)} + 4y''' + 6y'' + 4y' + y = 0$$

***A.*** $y = c_1 e^{-x} + c_2 x e^{-x} + c_3 x^2 e^{-x} + c_4 x^3 e^{-x}$

1. Here you have a fourth order differential equation with constant coefficients, so try a solution of the form

   $$y = e^{rx}$$

2. Plug your attempted solution into the differential equation:

   $$r^4 e^{rx} + 4r^3 e^{rx} + 6r^2 e^{rx} + 4r e^{rx} + e^{rx} = 0$$

3. Then divide by $e^{rx}$ to get the characteristic equation:

   $$r^4 + 4r^3 + 6r^2 + 4r + 1 = 0$$

4. Factor the characteristic equation as follows, either by hand or by using the equation-solving tool at www.quick math.com (see the chapter intro for specifics on accessing this tool):

   $$(r + 1)(r + 1)(r + 1)(r + 1)$$

5. The roots of the characteristic equation are $-1, -1, -1, -1$ — all repeated roots. Because the resulting solutions are all the same, these are all *degenerate solutions:*

   $$y_1 = e^{-x}$$
   $$y_2 = e^{-x}$$
   $$y_3 = e^{-x}$$
   $$y_4 = e^{-x}$$

6. Multiply the degenerate solutions by ascending powers of $x$ to get

   $$y_1 = c_1 e^{-x}$$
   $$y_2 = c_2 x e^{-x}$$
   $$y_3 = c_3 x^2 e^{-x}$$
   $$y_4 = c_4 x^3 e^{-x}$$

7. Put all four individual solutions together so that the general solution looks like

   $$y = c_1 e^{-x} + c_2 x e^{-x} + c_3 x^2 e^{-x} + c_4 x^3 e^{-x}$$

**9.** Obtain the solution to the following equation:

$$y''' + 3y'' + 3y' + y = 0$$

Solve It

**10.** Solve this differential equation:

$$y''' + 9y'' + 27y' + 27y = 0$$

Solve It

**11.** What's the solution to this equation?

$$y''' + 15y'' + 75y' + 125y = 0$$

Solve It

**12.** Obtain the solution to the following equation:

$$y''' + 4y'' + 5y' + 2y = 0$$

Solve It

**13.** Solve this differential equation:

$$y''' + 5y'' + 8y' + 4y = 0$$

Solve It

**14.** What's the solution to this equation?

$$y''' + 5y'' + 7y' + 3y = 0$$

Solve It

# Answers to Homogeneous Linear Higher Order Differential Equation Problems

Here are the answers to the practice questions I provide throughout this chapter. I walk you through each answer so you can see the problems worked out step by step. Enjoy!

**1** Find the solution to the following differential equation:

$$y''' + 7y'' + 14y' + 8y = 0$$

with these initial conditions:

$$y(0) = 3$$
$$y'(0) = -7$$
$$y''(0) = 14$$

**Solution:** $y = e^{-x} + e^{-2x} + e^{-4x}$

1. Because this differential equation has constant coefficients, start by assuming a solution of the following form:

$$y = e^{rx}$$

2. Plug your attempted solution into the differential equation:

$$r^3 e^{rx} + 7r^2 e^{rx} + 14r e^{rx} + 8 e^{rx} = 0$$

3. Then cancel out $e^{rx}$:

$$r^3 + 7r^2 + 14r + 8 = 0$$

4. Now you have a cubic equation, which you can factor into the following either by hand or by using the equation-solving tool at www.quickmath.com (see the chapter intro for specifics on accessing this tool):

$$(r + 1)(r + 2)(r + 4) = 0$$

5. So the roots are

$$r_1 = -1, r_2 = -2, \text{ and } r_3 = -4$$

6. These roots are real and distinct, so the solutions are

$$y_1 = e^{-x}, y_2 = e^{-2x}, \text{ and } y_3 = e^{-4x}$$

7. Thus, the general solution is

$$y = c_1 e^{-x} + c_2 e^{-2x} + c_3 e^{-4x}$$

8. Now you can apply the initial conditions. But in addition to the form for $y$, you also need $y'$:

$$y' = -c_1 e^{-x} - 2c_2 e^{-2x} - 4c_3 e^{-4x}$$

and $y''$:

$$y'' = c_1 e^{-x} + 4c_2 e^{-2x} + 16c_3 e^{-4x}$$

9. From the initial conditions, your three simultaneous equations in $c_1$, $c_2$, and $c_3$ that you must solve to find the coefficients are

$$y(0) = c_1 + c_2 + c_3 = 3$$
$$y'(0) = -c_1 - 2c_2 - 4c_3 = -7$$
$$y''(0) = c_1 + 4c_2 + 16c_3 = 14$$

10. If you solve this system of three equations by hand (or by using the $3 \times 3$ system solver at `math.cowpi.com/systemsolver`), you should get

$$c_1 = 1, c_2 = 1, \text{ and } c_3 = 1$$

11. So the solution of the differential equation with initial conditions applied is

$$y = e^{-x} + e^{-2x} + e^{-4x}$$

**2** **Solve this equation:**

$$y''' + 8y'' + 19y' + 12y = 0$$

where

$$y(0) = 4$$
$$y'(0) = -12$$
$$y''(0) = 42$$

**Solution:** $y = e^{-x} + e^{-3x} + 2e^{-4x}$

1. Begin by assuming a solution of the form

$$y = e^{rx}$$

2. Plugging your attempted solution into the differential equation gives you

$$r^3 e^{rx} + 8r^2 e^{rx} + 19r e^{rx} + 12 e^{rx} = 0$$

3. Canceling out $e^{rx}$ leaves you with

$$r^3 + 8r^2 + 19r + 12 = 0$$

4. Looks like you now have a cubic equation that can be factored as follows:

$$(r + 1)(r + 3)(r + 4) = 0$$

which means the roots are

$$r_1 = -1, r_2 = -3, \text{ and } r_3 = -4$$

5. Because the roots are real and distinct, the solutions are

$$y_1 = e^{-x}, y_2 = e^{-3x}, \text{ and } y_3 = e^{-4x}$$

6. That means the general solution must be

$$y = c_1 e^{-x} + c_2 e^{-3x} + c_3 e^{-4x}$$

7. Great. Now you can apply the initial conditions, but make sure to first find $y'$ as

$$y' = -c_1 e^{-x} - 3c_2 e^{-3x} - 4c_3 e^{-4x}$$

and $y''$ as

$$y'' = c_1 e^{-x} + 9c_2 e^{-3x} + 16c_3 e^{-4x}$$

8. Thanks to the initial conditions, your three simultaneous equations in $c_1$, $c_2$, and $c_3$ (which you have to solve to find the coefficients) are

$$y(0) = c_1 + c_2 + c_3 = 4$$
$$y'(0) = -c_1 - 3c_2 - 4c_3 = -12$$
$$y''(0) = c_1 + 9c_2 + 16c_3 = 42$$

9. When you solve this system of three equations, you get the following:

$$c_1 = 1, c_2 = 1, \text{ and } c_3 = 2$$

10. Therefore, the solution of the original equation with initial conditions applied is

$$y = e^{-x} + e^{-3x} + 2e^{-4x}$$

**3** **Obtain the solution to this equation:**

$$y''' + 9y'' + 26y' + 24y = 0$$

**by using these initial conditions:**

$$y(0) = 5$$
$$y'(0) = -16$$
$$y''(0) = 54$$

**Solution:** $y = e^{-2x} + 2e^{-3x} + 2e^{-4x}$

1. Because this differential equation has constant coefficients, start by assuming a solution of the following form:

$$y = e^{rx}$$

2. Plug your attempted solution into the differential equation:

$$r^3 e^{rx} + 9r^2 e^{rx} + 26r e^{rx} + 24 e^{rx} = 0$$

3. Then cancel out $e^{rx}$:

$$r^3 + 9r^2 + 26r + 24 = 0$$

4. Now you have a cubic equation, which you can factor into

$$(r + 2)(r + 3)(r + 4) = 0$$

5. So the roots are

$$r_1 = -2, r_2 = -3, \text{ and } r_3 = -4$$

6. These roots are real and distinct, so the solutions are

$$y_1 = e^{-2x}, y_2 = e^{-3x}, \text{ and } y_3 = e^{-4x}$$

7. Thus, the general solution is

$$y = c_1 e^{-2x} + c_2 e^{-3x} + c_3 e^{-4x}$$

8. Now you can apply the initial conditions. But in addition to the form for $y$, you also need $y'$:

$$y' = -2c_1 e^{-2x} - 3c_2 e^{-3x} - 4c_3 e^{-4x}$$

and $y''$:

$$y'' = 4c_1 e^{-2x} + 9c_2 e^{-3x} + 16c_3 e^{-4x}$$

9. From the initial conditions, your three simultaneous equations in $c_1$, $c_2$, and $c_3$ that you must solve to find the coefficients are

$$y(0) = c_1 + c_2 + c_3 = 5$$
$$y'(0) = -2c_1 - 3c_2 - 4c_3 = -16$$
$$y''(0) = 4c_1 + 9c_2 + 16c_3 = 54$$

10. Solving this system of three equations gives you

$$c_1 = 1, c_2 = 2, \text{ and } c_3 = 2$$

11. So the solution of the differential equation with initial conditions applied is

$$y = e^{-2x} + 2e^{-3x} + 2e^{-4x}$$

**4** **Find the solution to the following differential equation:**

$$y''' + 10y'' + 31y' + 30y = 0$$

**with these initial conditions:**

$$y(0) = 6$$
$$y'(0) = -19$$
$$y''(0) = 71$$

**Solution:** $y = 3e^{-2x} + e^{-3x} + 2e^{-5x}$

1. Begin by assuming a solution of the form

$$y = e^{rx}$$

2. Plugging your attempted solution into the differential equation gives you

$$r^3 e^{rx} + 10r^2 e^{rx} + 31r e^{rx} + 30 e^{rx} = 0$$

3. Canceling out $e^{rx}$ leaves you with

$$r^3 + 10r^2 + 31r + 30 = 0$$

4. Looks like you now have a cubic equation that can be factored as follows:

$$(r + 2)(r + 3)(r + 5) = 0$$

which means the roots are

$$r_1 = -2, r_2 = -3, \text{ and } r_3 = -5$$

5. Because the roots are real and distinct, the solutions are

$$y_1 = e^{-2x}, y_2 = e^{-3x}, \text{ and } y_3 = e^{-5x}$$

6. That means the general solution must be

$$y = c_1 e^{-2x} + c_2 e^{-3x} + c_3 e^{-5x}$$

7. Great. Now you can apply the initial conditions, but make sure to first find $y'$ as

$$y' = -2c_1 e^{-2x} - 3c_2 e^{-3x} - 5c_3 e^{-5x}$$

and $y''$ as

$$y'' = 4c_1 e^{-2x} + 9c_2 e^{-3}x + 25c_3 e^{-5x}$$

8. Thanks to the initial conditions, your three simultaneous equations in $c_1$, $c_2$, and $c_3$ (which you have to solve to find the coefficients) are

$$y(0) = c_1 + c_2 + c_3 = 6$$

$$y'(0) = -2c_1 - 3c_2 - 5c_3 = -19$$

$$y''(0) = 4c_1 + 9c_2 + 25c_3 = 71$$

9. When you solve this system of three equations, you get the following:

$$c_1 = 3, c_2 = 1, \text{ and } c_3 = 2$$

10. Therefore, the solution of the original equation with initial conditions applied is

$$y = 3e^{-2x} + e^{-3x} + 2e^{-5x}$$

**5**  **Solve this equation:**

$$y''' + 11y'' + 36y' + 36y = 0$$

**where**

$$y(0) = 5$$

$$y'(0) = -17$$

$$y''(0) = 67$$

**Solution: $y = e^{-2x} + 3e^{-3x} + e^{-6x}$**

1. Because this differential equation has constant coefficients, start by assuming a solution of the following form:

$$y = e^{rx}$$

2. Plug your attempted solution into the differential equation:

$$r^3 e^{rx} + 11r^2 e^{rx} + 36re^{rx} + 36e^{rx} = 0$$

3. Then cancel out $e^{rx}$:

$$r^3 + 11r^2 + 36r + 36 = 0$$

4. Now you have a cubic equation, which you can factor into

$$(r + 2)(r + 3)(r + 6) = 0$$

5. So the roots are

$$r_1 = -2, r_2 = -3, \text{ and } r_3 = -6$$

6. These roots are real and distinct, so the solutions are

$$y_1 = e^{-2x}, y_2 = e^{-3x}, \text{ and } y_3 = e^{-6x}$$

7. Thus, the general solution is

$$y = c_1 e^{-2x} + c_2 e^{-3x} + c_3 e^{-6x}$$

8. Now you can apply the initial conditions. But in addition to the form for $y$, you also need $y'$:

$$y' = -2c_1 e^{-2x} - 3c_2 e^{-3x} - 6c_3 e^{-6x}$$

and $y''$:

$$y'' = 4c_1 e^{-2x} + 9c_2 e^{-3x} + 36c_3 e^{-6x}$$

9. From the initial conditions, your three simultaneous equations in $c_1$, $c_2$, and $c_3$ that you must solve to find the coefficients are

$$y(0) = c_1 + c_2 + c_3 = 5$$
$$y'(0) = -2c_1 - 3c_2 - 6c_3 = -17$$
$$y''(0) = 4c_1 + 9c_2 + 36c_3 = 67$$

10. Solving this system of three equations gives you

$$c_1 = 1, c_2 = 3, \text{ and } c_3 = 1$$

11. So the solution of the differential equation with initial conditions applied is

$$y = e^{-2x} + 3e^{-3x} + e^{-6x}$$

6  **Obtain the solution to this equation:**

$$y''' + 10y'' + 29y' + 20y = 0$$

**by using these initial conditions:**

$$y(0) = 7$$
$$y'(0) = -30$$
$$y''(0) = 142$$

**Solution:** $y = e^{-x} + e^{-4x} + 5e^{-5x}$

1. Begin by assuming a solution of the form

$$y = e^{rx}$$

2. Plugging your attempted solution into the differential equation gives you

$$r^3 e^{rx} + 10r^2 e^{rx} + 29r e^{rx} + 20 e^{rx} = 0$$

3. Canceling out $e^{rx}$ leaves you with

$$r^3 + 10r^2 + 29r + 20 = 0$$

4. Looks like you now have a cubic equation that can be factored as follows:

$$(r + 1)(r + 4)(r + 5) = 0$$

which means the roots are

$$r_1 = -1, r_2 = -4, \text{ and } r_3 = -5$$

5. Because the roots are real and distinct, the solutions are

$$y_1 = e^{-x}, y_2 = e^{-4x}, \text{ and } y_3 = e^{-5x}$$

6. That means the general solution must be

$$y = c_1 e^{-x} + c_2 e^{-4x} + c_3 e^{-5x}$$

7. Great. Now you can apply the initial conditions, but make sure to first find $y'$ as

$$y' = -c_1 e^{-x} - 4c_2 e^{-4x} - 5c_3 e^{-5x}$$

and $y''$ as

$$y'' = c_1 e^{-2x} + 16c_2 e^{-3x} + 25c_3 e^{-5x}$$

8. Thanks to the initial conditions, your three simultaneous equations in $c_1$, $c_2$, and $c_3$ (which you have to solve to find the coefficients) are

$$y(0) = c_1 + c_2 + c_3 = 7$$

$$y'(0) = -c_1 - 4c_2 - 5c_3 = -30$$

$$y''(0) = c_1 + 16c_2 + 25c_3 = 142$$

9. When you solve this system of three equations, you get the following:

$$c_1 = 1, c_2 = 1, \text{ and } c_3 = 5$$

10. Therefore, the solution of the original equation with initial conditions applied is

$$y = e^{-x} + e^{-4x} + 5e^{-5x}$$

**7** **What's the solution to the following differential equation?**

$$y^{(4)} - 16y = 0$$

**where**

$$y(0) = 3$$

$$y'(0) = 2$$

$$y''(0) = -4$$

$$y'''(0) = -24$$

**Solution: $y = e^{-2x} + 2 \cos 2x + 2 \sin 2x$**

1. Because this differential equation has constant coefficients, you can safely assume a solution of the form

$$y = e^{rx}$$

2. Plug your solution into the equation to get

$$r^4 e^{rx} - 16 e^{rx} = 0$$

3. Then cancel out $e^{rx}$ to get

$$r^4 - 16 = 0$$

4. You can factor the resulting characteristic equation into

$$(r^2 - 4)(r^2 + 4) = 0$$

5. Looks like the roots of the characteristic equation are

$$r_1 = 2, r_2 = -2, r_3 = 2i, \text{ and } r_4 = -2i$$

6. Use these relationships to simplify:

$$e^{i\beta x} = \cos \beta x + i \sin \beta x$$

and

$$e^{-i\beta x} = \cos \beta x - i \sin \beta x$$

7. Well, $y_3$ and $y_4$ can be expressed as a linear combination of sines and cosines, which gives you these solutions (note that $i$ has been absorbed into a multiplicative constant):

$$y_1 = e^{2x}$$

$$y_2 = e^{-2x}$$

$$y_3 = \cos 2x$$

$$y_4 = \sin 2x$$

8. You can therefore determine that the general solution is

$$y = c_1 e^{2x} + c_2 e^{-2x} + c_3 \cos 2x + c_4 \sin 2x$$

9. To apply the initial conditions, you need to figure out $y'$:

$$y' = 2c_1 e^{2x} - 2c_2 e^{-2x} - 2c_3 \sin 2x + 2c_4 \cos 2x$$

$y''$:

$$y'' = 4c_1 e^{2x} + 4c_2 e^{-2x} - 4c_3 \cos 2x - 4c_4 \sin 2x$$

and $y'''$:

$$y''' = 8c_1 e^{2x} - 8c_2 e^{-2x} + 8c_3 \sin 2x - 8c_4 \cos 2x$$

10. Substituting the forms for $y$, $y'$, $y''$, and $y'''$ into the initial conditions gives you

$$y(0) = c_1 + c_2 + c_3 = 3$$
$$y'(0) = 2c_1 - 2c_2 + 2c_4 = 2$$
$$y''(0) = 4c_1 + 4c_2 - 4c_3 = -4$$
$$y'''(0) = 8c_1 - 8c_2 - 8c_4 = -24$$

11. Solve this $4 \times 4$ simultaneous equation system by hand or with an online tool such as the $4 \times 4$ system solver at `math.cowpi.com/systemsolver` to get the following:

$$c_1 = 0, c_2 = 1, c_3 = 2, \text{ and } c_4 = 2$$

12. Tada! The general solution with initial conditions applied is

$$y = e^{-2x} + 2 \cos 2x + 2 \sin 2x$$

**8** Find the solution to this differential equation:

$$y^{(4)} - 81y = 0$$

where

$$y(0) = 4$$

$$y'(0) = 12$$

$$y''(0) = -18$$

$$y'''(0) = -162$$

**Solution:** $y = e^{-3x} + 3 \cos 3x + 5 \sin 3x$

1. You know this differential equation has constant coefficients, so go ahead and assume a solution of this form:

$$y = e^{rx}$$

2. Plug your solution into the equation:

   $r^4 e^{rx} - 81 e^{rx} = 0$

3. Next, cancel out $e^{rx}$:

   $r^4 - 81 = 0$

4. You can factor the resulting characteristic equation into

   $(r^2 - 3)(r^2 + 3) = 0$

5. So the roots of the characteristic equation are

   $r_1 = 3, r_2 = -3, r_3 = 3i,$ and $r_4 = -3i$

6. To simplify, use these relationships:

   $e^{i\beta x} = \cos \beta x + i \sin \beta x$

   and

   $e^{-i\beta x} = \cos \beta x - i \sin \beta x$

7. Okay, so you've determined that $y_3$ and $y_4$ can be expressed as a linear combination of sines and cosines (note that you can absorb $i$ into a multiplicative constant). That means you have these solutions:

   $y_1 = e^{3x}$

   $y_2 = e^{-3x}$

   $y_3 = \cos 3x$

   $y_4 = \sin 3x$

8. So the general solution is

   $y = c_1 e^{3x} + c_2 e^{-3x} + c_3 \cos 3x + c_4 \sin 3x$

9. To apply the initial conditions, you need to figure out $y'$:

   $y' = 3c_1 e^{3x} - 3c_2 e^{-3x} - 3c_3 \sin 3x + 3c_4 \cos 3x$

   and $y''$:

   $y'' = 9c_1 e^{3x} + 9c_2 e^{-3x} - 9c_3 \cos 3x - 9c_4 \sin 3x$

   as well as $y'''$:

   $y''' = 27c_1 e^{3x} - 27c_2 e^{-3x} + 27c_3 \sin 3x - 27c_4 \cos 3x$

10. Substituting the forms for $y$, $y'$, $y''$, and $y'''$ into the initial conditions gives you

    $y(0) = c_1 + c_2 + c_3 = 4$

    $y'(0) = 3c_1 - 3c_2 + 3c_4 = 12$

    $y''(0) = 9c_1 + 9c_2 - 9c_3 = -18$

    $y'''(0) = 27c_1 - 27c_2 - 27c_4 = -162$

11. Solve this $4 \times 4$ simultaneous equation system by hand or with an online tool such as the $4 \times 4$ system solver at `math.cowpi.com/systemsolver`:

    $c_1 = 0, c_2 = 1, c_3 = 3,$ and $c_4 = 5$

12. After all that, the general solution with initial conditions applied is

$$y = e^{-3x} + 3\cos 3x + 5\sin 3x$$

**9** **Obtain the solution to the following equation:**

$$y''' + 3y'' + 3y' + y = 0$$

**Solution:** $y = c_1 e^{-x} + c_2 x e^{-x} + c_3 x^2 e^{-x}$

1. First, try a solution of the following form:

$$y = e^{rx}$$

2. Plug your attempted solution into the differential equation to get

$$r^3 e^{rx} + 3r^2 e^{rx} + 3r e^{rx} + e^{rx} = 0$$

3. Then divide by $e^{rx}$ to get this characteristic equation:

$$r^3 + 3r^2 + 3r + 1 = 0$$

4. Factor the characteristic equation either by hand or by using the equation-solving tool at www.quickmath.com (see the chapter intro for specifics on accessing this tool):

$$(r + 1)(r + 1)(r + 1)$$

5. It appears the roots of the characteristic equation are repeated roots, because all three roots are –1. Consequently, your three resulting solutions are considered degenerate solutions, because they're all the same.

$$y_1 = e^{-x}$$

$$y_2 = e^{-x}$$

$$y_3 = e^{-x}$$

6. Go ahead and multiply the degenerate solutions by ascending powers of $x$ to get

$$y_1 = c_1 e^{-x}$$

$$y_2 = c_2 x e^{-x}$$

$$y_3 = c_3 x^2 e^{-x}$$

7. Put the individual solutions together to form your general solution, like so:

$$y = c_1 e^{-x} + c_2 x e^{-x} + c_3 x^2 e^{-x}$$

**10** **Solve this differential equation:**

$$y''' + 9y'' + 27y' + 27y = 0$$

**Solution:** $y = c_1 e^{-3x} + c_2 x e^{-3x} + c_3 x_2 e^{-3x}$

1. Because the problem features a third order differential equation with constant coefficients, try a solution of the form

$$y = e^{rx}$$

2. Then plug the attempted solution into the equation:

$$r^3 e^{rx} + 9r^2 e^{rx} + 27r e^{rx} + 27 e^{rx} = 0$$

3. Divide by $e^{rx}$ to get the characteristic equation:

   $$r^3 + 9r^2 + 27r + 27 = 0$$

   which you can factor as follows:

   $$(r + 3)(r + 3)(r + 3)$$

4. The roots of the characteristic equation are –3, –3, –3 — all repeated roots — which give you these degenerate solutions:

   $$y_1 = e^{-3x}$$

   $$y_2 = e^{-3x}$$

   $$y_3 = e^{-3x}$$

5. Multiply the degenerate solutions by ascending powers of $x$:

   $$y_1 = c_1 e^{-3x},\ y_2 = c_2 x e^{-3x},\ \text{and}\ y_3 = c_3 x^2 e^{-3x}$$

6. Finally, put the individual solutions together to form this general solution:

   $$y = c_1 e^{-3x} + c_2 x e^{-3x} + c_3 x^2 e^{-3x}$$

**11** **What's the solution to this equation?**

$$y''' + 15y'' + 75y' + 125y = 0$$

**Solution:** $y = c_1 e^{-5x} + c_2 x e^{-5x} + c_3 x^2 e^{-5x}$

1. First, try a solution of the following form:

   $$y = e^{rx}$$

2. Plug your attempted solution into the differential equation to get

   $$r^3 e^{rx} + 15r^2 e^{rx} + 75r e^{rx} + 125 e^{rx} = 0$$

3. Then divide by $e^{rx}$ to get this characteristic equation:

   $$r^3 + 15r^2 + 75r + 125 = 0$$

4. Factor the characteristic equation this way:

   $$(r + 5)(r + 5)(r + 5)$$

5. It appears the roots of the characteristic equation are repeated roots, because all three roots are –5. Consequently, your three resulting solutions are considered degenerate solutions, because they're all the same.

   $$y_1 = e^{-5x}$$

   $$y_2 = e^{-5x}$$

   $$y_3 = e^{-5x}$$

6. Go ahead and multiply the degenerate solutions by ascending powers of $x$ to get

   $$y_1 = c_1 e^{-5x}$$

   $$y_2 = c_2 x e^{-5x}$$

   $$y_3 = c_3 x^2 e^{-5x}$$

7. Put the individual solutions together to form your general solution, like so:

$$y = c_1 e^{-5x} + c_2 x e^{-5x} + c_3 x^2 e^{-5x}$$

**12** **Obtain the solution to the following equation:**

$$y''' + 4y'' + 5y' + 2y = 0$$

**Solution:** $y = c_1 e^{-x} + c_2 x e^{-x} + c_3 e^{-2x}$

1. Because the problem features a third order differential equation with constant coefficients, try a solution of the form

$$y = e^{rx}$$

2. Then plug the attempted solution into the equation:

$$r^3 e^{rx} + 4r^2 e^{rx} + 5r e^{rx} + 2e^{rx} = 0$$

3. Divide by $e^{rx}$ to get the characteristic equation:

$$r^3 + 4r^2 + 5r + 2 = 0$$

which you can factor as follows:

$$(r + 1)(r + 1)(r + 2)$$

4. The roots of the characteristic equation are $-1, -1, -2$ — two of the roots are repeated. Find the solutions:

$$y_1 = e^{-x}$$

$$y_2 = e^{-x}$$

$$y_3 = e^{-2x}$$

5. Multiply the degenerate solutions by ascending powers of $x$:

$$y_1 = c_1 e^{-x}, y_2 = c_2 x e^{-x}, \text{ and } y_3 = c_3 e^{-2x}$$

6. Finally, put the individual solutions together to form this general solution:

$$y = c_1 e^{-x} + c_2 x e^{-x} + c_3 e^{-2x}$$

**13** **Solve this differential equation:**

$$y''' + 5y'' + 8y' + 4y = 0$$

**Solution:** $y = c_1 e^{-2x} + c_2 x e^{-2x} + c_3 e^{-x}$

1. First, try a solution of the following form:

$$y = e^{rx}$$

2. Plug your attempted solution into the differential equation to get

$$r^3 e^{rx} + 5r^2 e^{rx} + 8r e^{rx} + 4e^{rx} = 0$$

3. Then divide by $e^{rx}$ to get this characteristic equation:

$$r^3 + 5r^2 + 8r + 4 = 0$$

4. Factor the characteristic equation this way:

$$(r + 2)(r + 2)(r + 1)$$

5. It appears two of the characteristic equation's roots are repeated, –2 and –2. Consequently, two of your three resulting solutions are considered degenerate solutions, because they're the same.

$$y_1 = e^{-2x}$$

$$y_2 = e^{-2x}$$

$$y_3 = e^{-x}$$

6. Go ahead and multiply the degenerate solutions by ascending powers of $x$ to get

$$y_1 = c_1 e^{-2x}$$

$$y_2 = c_2 x e^{-2x}$$

$$y_3 = c_3 e^{-x}$$

7. Put the individual solutions together to form your general solution, like so:

$$y = c_1 e^{-2x} + c_2 x e^{-2x} + c_3 e^{-x}$$

**14** **What's the solution to this equation?**

$$y''' + 5y'' + 7y' + 3y = 0$$

**Solution:** $y = c_1 e^{-x} + c_2 x e^{-x} + c_3 e^{-3x}$

1. Because the problem features a third order differential equation with constant coefficients, try a solution of the form

$$y = e^{rx}$$

2. Then plug your attempted solution into the equation:

$$r^3 e^{rx} + 5r^2 e^{rx} + 7r e^{rx} + 3e^{rx} = 0$$

3. Divide by $e^{rx}$ to get the characteristic equation:

$$r^3 + 5r^2 + 7r + 3 = 0$$

which you can factor as follows:

$$(r + 1)(r + 1)(r + 3)$$

4. The roots of the characteristic equation are –1, –1, –3 — two of the roots are repeated. Find the solutions:

$$y_1 = e^{-x}$$

$$y_2 = e^{-x}$$

$$y_3 = e^{-3x}$$

5. Multiply the degenerate solutions by ascending powers of $x$:

$$y_1 = c_1 e^{-x}$$

$$y_2 = c_2 x e^{-x}$$

$$y_3 = c_3 e^{-3x}$$

6. Finally, put the individual solutions together to form this general solution:

$$y = c_1 e^{-x} + c_2 x e^{-x} + c_3 e^{-3x}$$

# Chapter 7

# Taking On Nonhomogeneous Linear Higher Order Differential Equations

## In This Chapter

▶ Finding answers with the help of $Ae^{rx}$

▶ Working with polynomial differential equations

▶ Knowing what to do when you see sines and cosines

**I**n this chapter, you work with general nonhomogeneous linear higher order differential equations (which are sometimes referred to as $n$th order equations) that look like this:

$$\frac{d^n y}{dx^n} + p_1(x)\frac{d^{n-1}y}{dx^{n-1}} + p_2(x)\frac{d^{n-2}y}{dx^{n-2}} + \dots + p_{n-1}(x)\frac{dy}{dx} + p_n(x)y = g(x)$$

Such an equation may seem complex, but you can easily solve it by using the method of undetermined coefficients for nonhomogeneous higher order differential equations.

The *method of undetermined coefficients* says that if $g(x)$ has a certain form, then you must attempt to find a particular solution of a similar form. After you find the particular solution, you must solve for the *general* solution, which is the sum of the homogeneous solution (which you find by setting $g(x)$ to 0) and the particular solution.

The various forms of $g(x)$ give you a clue as to what the form of the particular solution may be. If $g(x)$ equals

▶ $e^{rx}$, then try a particular solution of the form $Ae^{rx}$, where $A$ is a constant. Because derivatives of $e^{rx}$ reproduce $e^{rx}$, you have a good chance of finding a particular solution this way.

▶ **a polynomial of order $n$,** then try a polynomial of order $n$.

▶ **a combination of sines and cosines,** $\sin \alpha x + \cos \beta x$, then try a combination of sines and cosines with undetermined coefficients ($A \sin \beta x + B \cos \beta x$), plug into the differential equation, and solve for $A$ and $B$.

In this chapter, you practice working with each of these forms.

As you solve the practice problems throughout this chapter, you may need help factoring the characteristic equations that crop up in the process of finding the homogeneous solution. I recommend turning to a trusty Web-based equation solver, such as www.numberempire. com/equationsolver.php. (Just be sure to use "x^3" for $x^3$).

# Seeking Out Solutions of the Form Ae^{rx}

Solving for a solution in the form of $Ae^{rx}$ is what you should try first when you want to find a solution to the homogeneous version of a nonhomogeneous linear higher order equation, like this one:

$$y''' + 3y'' + 3y' + y = 432e^{5x}$$

You know that the general solution of this equation is the sum of the particular solution and the homogeneous solution. You also know that in variable-speak, the general solution looks like this:

$$y = y_h + y_p$$

Perfect. So now you need to get the homogeneous version of the previous differential equation, which is

$$y''' + 3y'' + 3y' + y = 0$$

Solve the homogeneous equation first; then plug in a particular solution of the form

$$y_p = Ae^{5x}$$

All that's left to do is solve for $A$!

That was just a quick outline of the process; the following example walks you through the steps of solving the previous equation from start to finish. Check it out and then take a shot at solving the related practice problems.

**Q.** Find the solution to this differential equation:

$$y''' + 3y'' + 3y' + y = 432e^{5x}$$

**A.** $y = c_1e^{-x} + c_2xe^{-x} + c_3x^2e^{-x} + 2e^{5x}$

1. Start by realizing that obtaining the general solution to the problem means finding the sum of the particular solution and the solution to the homogeneous version of the differential equation:

$$y = y_h + y_p$$

2. Then find the homogeneous version of the differential equation:

$$y''' + 3y'' + 3y' + y = 0$$

3. The homogeneous version has constant coefficients, so you can assume a homogeneous solution of the form

$$y = e^{rx}$$

4. Plugging your attempted solution into the differential equation gives you

$$r^3e^{rx} + 3r^2e^{rx} + 3re^{rx} + e^{rx} = 0$$

5. Canceling out $e^{rx}$ leaves you with

$$r^3 + 3r^2 + 3r + 1 = 0$$

6. Now you have a cubic equation, which you can factor into the following either by hand or by using the equation-solving tool at www.numberempire.com/equationsolver.php:

$$(r + 1)(r + 1)(r + 1) = 0$$

7. So the roots of the characteristic equation are

$$r_1 = -1, r_2 = -1, \text{ and } r_3 = -1$$

which gives you

$$y_1 = c_1 e^{-x}$$

$$y_2 = c_2 x e^{-x}$$

$$y_3 = c_3 x^2 e^{-x}$$

8. So the solution to the homogeneous differential equation is

$$y_h = c_1 e^{-x} + c_2 x e^{-x} + c_3 x^2 e^{-x}$$

9. Now you need to find a particular solution to the differential equation by using the method of undetermined coefficients. Start by assuming a solution of the form

$$y_p = A e^{5x}$$

10. Substitute $y_p$ into the differential equation to get

$$125 A e^{5x} + 75 A e^{5x} + 15 A e^{5x} + A e^{5x} = 432 e^{5x}$$

11. Then cancel out $e^{5x}$:

$$125A + 75A + 15A + A = 432$$

which is actually

$$216A = 432$$

so

$$A = 2$$

and the particular solution is

$$y_p = 2 e^{5x}$$

12. Add the homogeneous solution and the particular solution together to find that the general solution to the original differential equation is

$$y = c_1 e^{-x} + c_2 x e^{-x} + c_3 x^2 e^{-x} + 2 e^{5x}$$

---

**1.** What's the solution to this differential equation?

$$y''' - 6y'' + 11y' - 6y = 18 e^{4x}$$

*Solve It*

**2.** Solve the following equation:

$$y''' + 7y'' + 14y' + 8y = 378 e^{5x}$$

*Solve It*

**3.** Find the answer to this equation:

$$y''' + 8y'' + 19y' + 14y = -36e^{-5x}$$

Solve It

**4.** What's the solution to this differential equation?

$$y''' + 9y'' + 26y' + 24y = 12e^{-x}$$

Solve It

**5.** Solve the following equation:

$$y''' - 6y'' + 11y' - 6y = 48e^{5x}$$

Solve It

**6.** Find the answer to this equation:

$$y''' + 4y'' + 5y' + 2y = 68e^{-3x}$$

Solve It

# *Trying for a Solution in Polynomial Form*

Whenever you come across a linear higher order differential equation that's nonhomogeneous *and* in polynomial form, forget the other tricks to the method of undetermined coefficients and try for a polynomial of order $n$.

With that in mind, how would you handle this equation?

$$y''' + 3y'' + 3y' + y = x + 5$$

Obviously the general solution you need to find is the sum of the homogeneous solution and a particular solution. The formula for that looks like this:

$$y = y_h + y_p$$

The homogeneous version of the original differential equation is

$$y''' + 3y'' + 3y' + y = 0$$

Okay. Now what? Well, first you must solve this homogeneous equation and then plug in a particular solution of the form

$$y_p = Ax^4 + Bx^3 + Cx^2 + Dx + E$$

and solve for $A$, $B$, $C$, $D$, and $E$. Nothing to it, right? For the step-by-step process, take a look at the following example. Or if you think you have the hang of it from this general overview, skip ahead to the following two practice problems.

**Q.** What's the solution to this differential equation?

$$y''' + 3y'' + 3y' + y = x + 5$$

**A.** $y = c_1 e^{-x} + c_2 x e^{-x} + c_3 x^2 e^{-x} + x + 2$

1. First things first: Make sure you're looking for the sum of the particular solution and the solution to the homogeneous version of the differential equation:

$$y = y_h + y_p$$

2. Now you can find the homogeneous version of the equation in the question:

$$y''' + 3y'' + 3y' + y = 0$$

3. Because the homogeneous differential equation has constant coefficients, go ahead and assume a homogeneous solution of the form

$$y = e^{rx}$$

4. Plug your attempted solution into the equation:

$$r^3 e^{rx} + 3r^2 e^{rx} + 3r e^{rx} + e^{rx} = 0$$

5. Then cancel out $e^{rx}$:

$$r^3 + 3r^2 + 3r + 1 = 0$$

6. The resulting equation is a cubic one that you can factor as follows either by hand or by using the equation-solving tool found at www.numberempire. com/equationsolver.php:

$$(r+1)(r+1)(r+1) = 0$$

7. This equation's roots are

$$r_1 = -1, r_2 = -1, \text{ and } r_3 = -1$$

so

$$y_1 = c_1 e^{-x}$$
$$y_2 = c_2 x e^{-x}$$
$$y_3 = c_3 x^2 e^{-x}$$

8. Tada! The solution to the homogeneous differential equation is

$$y_h = c_1 e^{-x} + c_2 x e^{-x} + c_3 x^2 e^{-x}$$

9. To find a particular solution to the differential equation, assume a solution of the form

$$y_p = Ax^4 + Bx^3 + Cx^2 + Dx + E$$

10. First, find $y_p'''$:

$$y_p''' = 24Ax + 6B$$

11. Then find $3y_p''$:

$$3y_p'' = 36Ax^2 + 18Bx + 6C$$

12. Next, find $3y_p'$:

$$3y_p' = 12Ax^3 + 9Bx^2 + 6Cx + 3D$$

13. Finally, find $y_p$:

$$y_p = Ax^4 + Bx^3 + Cx^2 + Dx + E$$

14. Add together the results of Steps 10 through 13:

$$y''' + 3y'' + 3y' + y = 24Ax + 6B + 36Ax^2 + 18Bx + 6C + 12Ax^3 + 9Bx^2 + 6Cx + 3D + Ax^4 + Bx^3 + Cx^2 + Dx + E = x + 5$$

15. That's a lot of terms to mess with. Why not combine them to get

$$Ax^4 + (12A + B)x^3 + (36A + 9B + C)x^2 + (24A + 18B + 6C + D)x + (6B + 6C + 3D + E) = x + 5$$

16. Comparing the coefficient of $x^4$ gives you

$$A = 0$$

17. Similarly, comparing the coefficient of $x^3$ gives you

$$12A + B = 0$$

so

$$B = 0$$

18. Amazingly enough, comparing the coefficient of $x^2$ gives you

$$36A + 9B + C = 0$$

so

$$C = 0$$

19. And comparing the coefficient of $x$ gives you

$$24A + 18B + 6C + D = 1$$

so

$$D = 1$$

20. Finally, comparing the coefficient of the constant term gives you

$$6B + 6C + 3D + E = 5$$

so

$$E = 2$$

21. After all that work, you now have your particular solution, which is

$$y_p = x + 2$$

22. Add that solution to the homogeneous solution and you get this as your general solution:

$$y = c_1 e^{-x} + c_2 x e^{-x} + c_3 x^2 e^{-x} + x + 2$$

**7.** Find the solution to the following differential equation:

$$y''' + 9y'' + 26y' + 24y = 24x + 2$$

*Solve It*

**8.** What's the solution to this equation?

$$y''' + 4y'' + 5y' + 2y = 4x + 16$$

*Solve It*

# Working with Solutions Made Up of Sines and Cosines

Spotting a sine or cosine in a nonhomogeneous linear higher order differential equation you're facing is a surefire sign that you need to find a particular solution that includes sines and cosines so you can plug it into the equation and solve for $A$ and $B$.

Say you're tackling this differential equation:

$$y''' + 7y'' + 14y' + 8y = 5 \sin (x)$$

You're well aware that the general solution is of this form:

$$y = y_h + y_p$$

where $y_h$ equals the homogeneous solution and $y_p$ equals the particular solution.

The homogeneous version of your original equation is

$$y''' + 7y'' + 14y' + 8y = 0$$

What you need to do next is solve this homogeneous equation first and then plug in a particular solution of the form

$$y_p = A \sin(x) + B \cos(x)$$

so you can solve for $A$ and $B$.

The following example shows you how to solve this type of equation. Spend a few minutes reviewing it before trying out a couple practice problems on your own.

**Q.** Find the solution to this differential equation:

$$y''' + 7y'' + 14y' + 8y = 2 \sin(x) + 26 \cos(x)$$

**A.** $y = c_1 e^{-x} + c_2 e^{-2x} + c_3 e^{-4x} + 2 \sin(x)$

1. You already know that the general solution looks like this:

$$y = y_h + y_p$$

so start by finding the homogeneous version of the differential equation:

$$y''' + 7y'' + 14y' + 8y = 0$$

2. You can assume a homogeneous solution of the following form because the homogeneous version of the equation has constant coefficients:

$$y = e^{rx}$$

3. Plug in your attempted solution to get

$$r^3 e^{rx} + 7r^2 e^{rx} + 14r e^{rx} + 8 e^{rx} = 0$$

4. Then cancel out $e^{rx}$:

$$r^3 + 7r^2 + 14r + 8 = 0$$

5. Looks like you now have a cubic equation on your hands. Don't worry. This one isn't as scary as it may look. Simply factor it as follows (either by hand or with the help of the equation-solving tool at www.numberempire.com/equationsolver.php):

$$(r + 1)(r + 2)(r + 4) = 0$$

6. Doing so shows you that the roots are

$$r_1 = -1, r_2 = -2, \text{ and } r_3 = -4$$

so

$$y_1 = e^{-x}$$
$$y_2 = e^{-2x}$$
$$y_3 = e^{-4x}$$

7. Thus, your homogeneous solution is

$$y_h = c_1 e^{-x} + c_2 e^{-2x} + c_3 e^{-4x}$$

8. Now you need to find a particular solution. Start that process by assuming a solution of this form:

$$y_p = A \sin(x) + B \cos(x)$$

9. Find the following:

$$y_p'''$$

which is

$$y_p''' = -A \cos(x) + B \sin(x)$$

then

$$7y_p''$$

which is

$$7y_p'' = -7A \sin(x) - 7B \cos(x)$$

then

$$14y_p'$$

which is

$$14y_p' = 14A \cos(x) - 14B \sin(x)$$

and finally

$$8y_p$$

which is

$$8y_p = 8A \sin(x) + 8B \cos(x)$$

10. Adding everything together gives you:

$$y''' + 7y'' + 14y' + 8y = -A \cos(x) + B \sin(x) - 7A \sin(x) - 7B \cos(x) + 14A \cos(x) - 14B \sin(x) + 8A \sin(x) + 8B \cos(x) = 5 \sin(x)$$

11. To get rid of some of the mess, go ahead and combine terms:

$$(B - 7A - 14B + 8A) \sin(x) + (-A - 7B + 14A + 8B) \cos(x) = 5 \sin(x)$$

12. Combining terms further gives you

$$(-13B + A) \sin(x) + (13A + B) \cos(x) = 2 \sin(x) + 26 \cos(x)$$

13. When you solve for $A$, you get

$$A = 2$$

and when you solve for $B$, you get

$$B = 0$$

so the particular solution is

$$y_p = 2 \sin(x)$$

14. Only one thing left to do: Add the particular solution to the homogeneous solution to get

$$y = c_1 e^{-x} + c_2 e^{-2x} + c_3 e^{-4x} + 2 \sin(x)$$

---

**9.** Solve the following differential equation:

$$y''' + 10y'' + 31y' + 30y = 20 \sin(x) + 30 \cos(x)$$

*Solve It*

**10.** What's the answer to this equation?

$$y''' + 11y'' + 36y' + 36y = 75 \sin(x) + 105 \cos(x)$$

*Solve It*

# Answers to Nonhomogeneous Linear Higher Order Differential Equation Problems

Following are the answers to the practice questions presented throughout this chapter. Each one is worked out step by step so that if you messed one up along the way, you can more easily see where you took a wrong turn.

**_1_** **What's the solution to this differential equation?**

$$y''' - 6y'' + 11y' - 6y = 18e^{4x}$$

**Solution:** $y = c_1 e^x + c_2 e^{2x} + c_3 e^{3x} + 3e^{4x}$

1. First off, get the homogeneous version of the equation:

   $$y''' - 6y'' + 11y' - 6y = 0$$

2. Looks like the homogeneous version has constant coefficients, so go ahead and assume a homogeneous solution of the form

   $$y = e^{rx}$$

3. Plug your attempted solution into the differential equation:

   $$r^3 e^{rx} - 6r^2 e^{rx} + 11r e^{rx} - 6e^{rx} = 0$$

4. Cancel out $e^{rx}$:

   $$r^3 - 6r^2 + 11r - 6 = 0$$

5. Now you have a cubic equation that can be factored into

   $$(r - 1)(r - 2)(r - 3) = 0$$

   **_Note:_** For some help factoring cubic (or higher!) equations, you can always turn to www.numberempire.com/equationsolver.php.

6. So the roots are

   $$r_1 = 1, r_2 = 2, \text{ and } r_3 = 3$$

7. Hmmm. Those roots are real and distinct, which makes the solutions

   $$y_1 = e^x$$
   $$y_2 = e^{2x}$$
   $$y_3 = e^{3x}$$

8. Therefore, you can calculate that the homogeneous solution is

   $$y_h = c_1 e^x + c_2 e^{2x} + c_3 e^{3x}$$

9. Well done. Now you need to assume a solution of the following form in order to find your particular solution:

   $$y_p = Ae^{4x}$$

10. Substitute $y_p$ into the differential equation to get

    $$64Ae^{4x} - 96Ae^{4x} + 44Ae^{4x} - 6Ae^{4x} = 18e^{4x}$$

11. Canceling out $e^{4x}$ gives you

   $$6A = 18$$

   which is actually

   $$A = 3$$

12. The particular solution is therefore

   $$y_p = 3e^{4x}$$

13. Add the particular solution to the homogeneous solution to get your general solution of

   $$y = c_1 e^x + c_2 e^{2x} + c_3 e^{3x} + 3e^{4x}$$

**2** **Solve the following equation:**

   $$y''' + 7y'' + 14y' + 8y = 378e^{5x}$$

**Solution:** $y = c_1 e^{-x} + c_2 e^{-2x} + c_3 e^{-4x} + e^{5x}$

1. Find the homogeneous version of the differential equation:

   $$y''' + 7y'' + 14y' + 8y = 0$$

2. Because the homogeneous version has constant coefficients, you can try a solution of the form

   $$y = e^{rx}$$

3. Plugging your attempted solution into the differential equation gives you

   $$r^3 e^{rx} + 7r^2 e^{rx} + 14r e^{rx} + 8e^{rx} = 0$$

4. Canceling out $e^{rx}$ leaves you with

   $$r^3 + 7r^2 + 14r + 8 = 0$$

5. Factor the resulting cubic equation as follows:

   $$(r + 1)(r + 2)(r + 4) = 0$$

6. You now know that the roots are

   $$r_1 = -1, r_2 = -2, \text{ and } r_3 = -4$$

7. The roots are real and distinct, so the solutions are

   $$y_1 = e^{-x}$$
   $$y_2 = e^{-2x}$$
   $$y_3 = e^{-4x}$$

8. Thus, the homogeneous solution must be

   $$y_h = c_1 e^{-x} + c_2 e^{-2x} + c_3 e^{-4x}$$

9. You still need to find a particular solution, so assume a solution of the form

   $$y_p = Ae^{5x}$$

10. Substituting $y_p$ into the differential equation gives you

   $$125Ae^{5x} + 175Ae^{5x} + 70Ae^{5x} + 8Ae^{5x} = 378e^{5x}$$

11. Canceling out $e^{5x}$ leaves you with the following:

$$125A + 175A + 70A + 8A = 378$$

or

$$378A = 378$$

which means

$$A = 1$$

12. So the particular solution is

$$y_p = e^{5x}$$

13. To get the general solution you're looking for, simply add together the homogeneous solution and the particular solution:

$$y = c_1e^{-x} + c_2e^{-2x} + c_3e^{-4x} + e^{5x}$$

**3** **Find the answer to this equation:**

$$y''' + 8y'' + 19y' + 14y = -36e^{-5x}$$

**Solution:** $y = c_1e^{-x} + c_2e^{-3x} + c_3e^{-4x} + 6e^{-5x}$

1. First off, get the homogeneous version of the equation:

$$y''' + 8y'' + 19y' + 14y = 0$$

2. Looks like the homogeneous version has constant coefficients, so go ahead and assume a solution of the form

$$y = e^{rx}$$

3. Plug your attempted solution into the differential equation:

$$r^3e^{rx} + 8r^2e^{rx} + 19re^{rx} + 14e^{rx} = 0$$

4. Cancel out $e^{rx}$:

$$r^3 + 8r^2 + 19r + 14 = 0$$

5. Now you have a cubic equation that can be factored into

$$(r + 1)(r + 3)(r + 4) = 0$$

6. So the roots are

$$r_1 = -1, r_2 = -3, \text{ and } r_3 = -4$$

7. Hmmm. Those roots are real and distinct, which makes the solutions

$$y_1 = e^{-x}$$
$$y_2 = e^{-3x}$$
$$y_3 = e^{-4x}$$

8. Therefore, you can calculate that the homogeneous solution is

$$y_h = c_1e^{-x} + c_2e^{-3x} + c_3e^{-4x}$$

9. Well done. Now you need to assume a solution of the following form in order to find your particular solution:

$$y_p = Ae^{-5x}$$

10. Substitute $y_p$ into the differential equation to get

$$-125Ae^{-5x} + 200Ae^{-5x} - 95Ae^{-5x} + 14Ae^{-5x} = -36e^{-5x}$$

11. Canceling out $e^{-5x}$ gives you

$$-125A + 200A - 95A + 14A = -36$$

which is actually

$$-6A = -36$$

so

$$A = 6$$

12. The particular solution is therefore

$$y_p = 6e^{-5x}$$

13. Add the particular solution to the homogeneous solution to get your general solution of

$$y = c_1e^{-x} + c_2e^{-3x} + c_3e^{-4x} + 6e^{-5x}$$

**4** **What's the solution to this differential equation?**

$$y''' + 9y'' + 26y' + 24y = 12e^{-x}$$

**Solution: $y = c_1e^{-2x} + c_2e^{-3x} + c_3e^{-4x} + 2e^{-x}$**

1. Find the homogeneous version of the differential equation:

$$y''' + 9y'' + 26y' + 24y = 0$$

2. Because the homogeneous version has constant coefficients, you can try a solution of the form

$$y = e^{rx}$$

3. Plugging your attempted solution into the differential equation gives you

$$r^3e^{rx} + 9r^2e^{rx} + 26re^{rx} + 24e^{rx} = 0$$

4. Canceling out $e^{rx}$ leaves you with

$$r^3 + 9r^2 + 26r + 24 = 0$$

5. Factor the resulting cubic equation as follows:

$$(r + 2)(r + 3)(r + 4) = 0$$

6. You now know that the roots are

$$r_1 = -2, r_2 = -3, \text{ and } r_3 = -4$$

7. The roots are real and distinct, so the solutions are

$$y_1 = e^{-2x}$$
$$y_2 = e^{-3x}$$
$$y_3 = e^{-4x}$$

8. Thus, the homogeneous solution must be

$$y_h = c_1e^{-2x} + c_2e^{-3x} + c_3e^{-4x}$$

9. You still need to find a particular solution, so assume a solution of the form

$$y_p = Ae^{-x}$$

10. Substituting $y_p$ into the differential equation gives you

$$-Ae^{-x} + 9Ae^{-x} - 26Ae^{-x} + 24Ae^{-x} = 12e^{-x}$$

11. Canceling out $e^{-x}$ leaves you with the following:

$$-A + 9A - 26A + 24A = 12$$

or

$$6A = 12$$

which means

$$A = 2$$

12. So the particular solution is

$$y_p = 2e^{-x}$$

13. To get the general solution you're looking for, simply add together the homogeneous solution and the particular solution:

$$y = c_1e^{-2x} + c_2e^{-3x} + c_3e^{-4x} + 2e^{-x}$$

**5** **Solve the following equation:**

$$y''' - 6y'' + 11y' - 6y = 48e^{5x}$$

**Solution:** $y = c_1e^x + c_2e^{2x} + c_3e^{3x} + 2e^{5x}$

1. First off, get the homogeneous version of the equation:

$$y''' - 6y'' + 11y' - 6y = 0$$

2. Looks like the homogeneous version has constant coefficients, so go ahead and assume a homogeneous solution of the form

$$y = e^{rx}$$

3. Plug your attempted solution into the differential equation:

$$r^3e^{rx} - 6r^2e^{rx} + 11re^{rx} - 6e^{rx} = 0$$

4. Cancel out $e^{rx}$:

$$r^3 - 6r^2 + 11r - 6 = 0$$

5. Now you have a cubic equation that can be factored into

$$(r-1)(r-2)(r-3) = 0$$

6. So the roots are

$$r_1 = 1, r_2 = 2, \text{ and } r_3 = 3$$

7. Hmmm. Those roots are real and distinct, which makes the solutions

$$y_1 = e^x$$
$$y_2 = e^{2x}$$
$$y_3 = e^{3x}$$

8. Therefore, you can calculate that the homogeneous solution is

$$y_h = c_1e^x + c_2e^{2x} + c_3e^{3x}$$

9. Well done. Now you need to assume a solution of the following form in order to find your particular solution:

$$y_p = Ae^{5x}$$

10. Substitute $y_p$ into the differential equation to get

$$125Ae^{5x} - 150Ae^{5x} + 55Ae^{5x} - 6Ae^{5x} = 48e^{5x}$$

11. Canceling out $e^{5x}$ gives you

$$24A = 48$$

which is actually

$$A = 2$$

12. The particular solution is therefore

$$y_p = 2e^{5x}$$

13. Add the particular solution to the homogeneous solution to get your general solution of

$$y = c_1 e^x + c_2 e^{2x} + c_3 e^{3x} + 2e^{5x}$$

**6**    **Find the answer to this equation:**

$$y''' + 4y'' + 5y' + 2y = 68e^{-3x}$$

**Solution:** $y = c_1 e^{-x} + c_2 x e^{-x} + c_3 e^{-2x} - 17e^{-3x}$

1. Find the homogeneous version of the differential equation:

$$y''' + 4y'' + 5y' + 2y = 0$$

2. Because the homogeneous version has constant coefficients, you can try a solution of the form

$$y = e^{rx}$$

3. Plugging your attempted solution into the differential equation gives you

$$r^3 e^{rx} + 4r^2 e^{rx} + 5r e^{rx} + 2e^{rx} = 0$$

4. Dividing by $e^{rx}$ to get the characteristic equation leaves you with

$$r^3 + 4r^2 + 5r + 2 = 0$$

5. Factor the characteristic equation as follows:

$$(r + 1)(r + 1)(r + 2)$$

6. Notice that two of the roots of the characteristic equation (–1 and –1) are repeated. Does it look as if the solutions are

$$y_1 = e^{-x}$$
$$y_2 = e^{-x}$$
$$y_3 = e^{-2x}$$

The first two are clearly degenerate solutions because they're the same.

7. Go ahead and multiply the degenerate solutions by ascending powers of $x$ to get

$$y_1 = c_1 e^{-x}$$
$$y_2 = c_2 x e^{-x}$$
$$y_3 = c_3 e^{-2x}$$

8. Put the individual solutions together so that the homogeneous solution looks like this:

$$y = c_1 e^{-x} + c_2 x e^{-x} + c_3 e^{-2x}$$

9. You still need to find a particular solution, so assume a solution of the form

$$y_p = Ae^{-3x}$$

10. Substituting $y_p$ into the differential equation gives you

$$-27Ae^{-3x} + 36Ae^{-3x} - 15Ae^{-3x} + 2Ae^{-3x} = 68e^{-3x}$$

11. Canceling out $e^{-3x}$ leaves you with the following:

$$-4A = 68$$

or

$$A = -17$$

12. So the particular solution is

$$y_p = -17e^{-3x}$$

13. To get the general solution you're looking for, simply add together the homogeneous solution and the particular solution:

$$y = c_1 e^{-x} + c_2 x e^{-x} + c_3 e^{-2x} - 17e^{-3x}$$

**7** **Find the solution to the following differential equation:**

$$y''' + 9y'' + 26y' + 24y = 24x + 2$$

**Solution:** $y = c_1 e^{-2x} + c_2 e^{-3x} + c_3 e^{-4x} + x - 1$

1. First things first: Obtain the homogeneous version of the differential equation:

$$y''' + 9y'' + 26y' + 24y = 0$$

2. Looks like the homogeneous version has constant coefficients, so you can safely try a solution of the following form:

$$y = e^{rx}$$

3. Now plug your attempted solution into the equation:

$$r^3 e^{rx} + 9r^2 e^{rx} + 26r e^{rx} + 24e^{rx} = 0$$

4. Then cancel out $e^{rx}$:

$$r^3 + 9r^2 + 26r + 24 = 0$$

5. You now have a cubic equation, which you can factor as follows either by hand or by using the equation-solving tool found at www.numberempire.com/equationsolver.php:

$$(r + 2)(r + 3)(r + 4) = 0$$

6. The roots are

$$r_1 = -2, r_2 = -3, \text{ and } r_3 = -4$$

so

$$y_1 = e^{-2x}$$

$$y_2 = e^{-3x}$$

$$y_3 = e^{-4x}$$

which makes the homogeneous solution

$$y_h = c_1 e^{-2x} + c_2 e^{-3x} + c_3 e^{-4x}$$

7. Now you need to find a particular solution to the differential equation by using the method of undetermined coefficients. Because $g(x)$ has the form of a polynomial in the original equation, start by assuming a solution of the form

$$y_p = Ax^4 + Bx^3 + Cx^2 + Dx + E$$

8. First, find $y_p'''$:

$$y_p''' = 24Ax + 6B$$

9. Then find $9y_p''$:

$$9y_p'' = 108Ax^2 + 54Bx + 18C$$

10. Next, find $26y_p'$:

$$26y_p' = 104Ax^3 + 78Bx^2 + 52Cx + 26D$$

11. Finally, find $24y_p$:

$$24y_p = 24Ax^4 + 24Bx^3 + 24Cx^2 + 24Dx + 24E$$

12. Add together the results of Steps 8 through 11:

$$y''' + 9y'' + 26y' + 24y = 24Ax + 6B + 108Ax^2 + 54Bx + 18C + 104Ax^3 + 78Bx^2 + 52Cx + 26D + 24Ax^4 + 24Bx^3 + 24Cx^2 + 24Dx + 24E = 24x + 2$$

13. Looks a wee bit nasty, doesn't it? Combining terms gives you

$$24Ax^4 + (104A + 24B)x^3 + (108A + 78B + 24C)x^2 + (24A + 54B + 52C + 24D)x + (6B + 18C + 26D + 24E) = 24x + 2$$

14. That's somewhat better. Try comparing the coefficient of $x^4$ to get

$$A = 0$$

15. Then compare the coefficient of $x^3$:

$$104A + 24B = 0$$

so

$$B = 0$$

16. While you're at it, why not compare the coefficient of $x^2$ to get

$$108A + 78B + 24C = 0$$

or

$$C = 0$$

17. Then compare the coefficient of $x$ (this is *almost* the last time you have to compare coefficients in this problem, I swear):

$$24A + 54B + 52C + 24D = 24$$

so

$$D = 1$$

18. Finally, compare the coefficient of the constant term to get

$$6B + 18C + 26D + 24E = 2$$

or

$E = -1$

19. If you put all that together, you can figure out that the particular solution is

$y_p = x - 1$

20. Take it one step further by adding the homogeneous solution and the particular solution to get your general solution of

$y = c_1 e^{-2x} + c_2 e^{-3x} + c_3 e^{-4x} + x - 1$

**8** **What's the solution to this equation?**

$y''' + 4y'' + 5y' + 2y = 4x + 16$

**Solution:** $y = c_1 e^{-x} + c_2 x e^{-x} + c_3 e^{-2x} + 2x + 3$

1. Find the homogeneous version of the equation in question:

$y''' + 4y'' + 5y' + 2y = 0$

2. Because the homogeneous version is a third order differential equation with constant coefficients, assume a solution of the form

$y = e^{rx}$

3. Plugging your attempted solution into the equation gives you

$r^3 e^{rx} + 4r^2 e^{rx} + 5r e^{rx} + 2e^{rx} = 0$

4. Dividing by $e^{rx}$ to get the characteristic equation results in this:

$r^3 + 4r^2 + 5r + 2 = 0$

5. Go ahead and factor the characteristic equation as

$(r + 1)(r + 1)(r + 2)$

6. Two of the roots of the characteristic equation are repeated roots (−1 and −1). Does it look as if the solutions are

$y_1 = e^{-x}$

$y_2 = e^{-x}$

$y_3 = e^{-2x}$

The first two solutions are degenerate because they're the same.

7. Multiply the degenerate solutions by ascending powers of $x$ to get

$y_1 = c_1 e^{-x}$

$y_2 = c_2 x e^{-x}$

$y_3 = c_3 e^{-2x}$

8. Then put the individual solutions together to form the following homogeneous solution:

$y = c_1 e^{-x} + c_2 x e^{-x} + c_3 e^{-2x}$

9. You're halfway there now. The next step is to find a particular solution, which you can do by assuming a solution of the form

$y_p = Ax^4 + Bx^3 + Cx^2 + Dx + E$

10. Find the following:

    $y_p'''$

    which is

    $y_p''' = 24Ax + 6B$

    then

    $4y_p''$

    which is

    $4y_p'' = 48Ax^2 + 34Bx + 8C$

    then

    $5y_p'$

    which is

    $5y_p' = 20Ax^3 + 15Bx^2 + 10Cx + 5D$

    and finally

    $2y_p$

    which is

    $2y_p = 2Ax^4 + 2Bx^3 + 2Cx^2 + 2Dx + 2E$

11. Add everything in Step 10 together:

    $y''' + 4y'' + 5y' + 2y = 24Ax + 6B + 48Ax^2 + 34Bx + 8C + 20Ax^3 + 15Bx^2 + 10Cx + 5D + 2Ax^4 + 2Bx^3 + 2Cx^2 + 2Dx + 2E = 4x + 16$

12. If that's a little messy for your taste, combine like terms:

    $2Ax^4 + (20A + 2B)x^3 + (48A + 15B + 2C)x^2 + (24A + 34B + 10C + 2D)x + (6B + 8C + 5D + 2E) = 4x + 16$

13. Now you can begin comparing coefficients, starting with the coefficient of $x^4$, which gives you

    $A = 0$

14. Next up, compare the coefficient of $x^3$ to get

    $20A + 2B = 0$

    so

    $B = 0$

15. Then compare the coefficient of $x^2$:

    $48A + 15B + 2C = 0$

    which means that

    $C = 0$

16. Comparing the coefficient of $x$ gives you

    $24A + 34B + 10C + 2D = 4$

    so

    $D = 2$

17. Last but not least, compare the coefficient of the constant term to get

$$6B + 8C + 5D + 2E = 16$$

which means that

$$E = 3$$

18. Thanks to all that work, you now know that the particular solution is

$$y_p = 2x + 3$$

19. Bring it home by finding the sum of the homogeneous solution and the particular solution (which gives you the general solution you were looking for to begin with!):

$$y = c_1 e^{-x} + c_2 x e^{-x} + c_3 e^{-2x} + 2x + 3$$

**9** **Solve the following differential equation:**

$$y''' + 10y'' + 31y' + 30y = 20 \sin(x) + 30 \cos(x)$$

**Solution: $y = c_1 e^{-2x} + c_2 e^{-3x} + c_3 e^{-5x} + \sin(x)$**

1. Your first step is to get the homogeneous version of the differential equation:

$$y''' + 10y'' + 31y' + 30y = 0$$

2. Try a solution of the following form because the homogeneous version of the equation has constant coefficients:

$$y = e^{rx}$$

3. Plug your attempted solution into the equation:

$$r^3 e^{rx} + 10r^2 e^{rx} + 31r e^{rx} + 30 e^{rx} = 0$$

4. Then cancel out $e^{rx}$ to get

$$r^3 + 10r^2 + 31r + 30 = 0$$

5. Factor the resulting cubic equation this way, either by hand or with the help of the equation-solving tool at www.numberempire.com/equationsolver.php:

$$(r + 2)(r + 3)(r + 5) = 0$$

6. Looks like the roots are

$$r_1 = -2, r_2 = -3, \text{ and } r_3 = -5$$

so

$$y_1 = e^{-2x}$$
$$y_2 = e^{-3x}$$
$$y_3 = e^{-5x}$$

7. Thus, the homogeneous solution is

$$y_h = c_1 e^{-2x} + c_2 e^{-3x} + c_3 e^{-5x}$$

8. Well done. Now you need to assume a solution of the following form in order to find your particular solution:

$$y_p = A \sin(x) + B \cos(x)$$

9. First, find $y_p'''$:

$$y_p''' = -A \cos(x) + B \sin(x)$$

10. Then find $10y_p''$:

$$10y_p'' = -10A \sin (x) - 10B \cos (x)$$

11. Next, find $31y_p'$:

$$31y_p' = 31A \cos (x) - 31B \sin (x)$$

12. Finally, find $30y_p$:

$$30y_p = 30A \sin (x) + 30B \cos (x)$$

13. Add together the results from Steps 9 through 12:

$$y''' + 10y'' + 31y' + 30y = -A \cos (x) + B \sin (x) - 10A \sin (x) - 10B \cos (x) + 31A \cos (x) - 31B \sin (x) + 30A \sin (x) + 30B \cos (x) = 20 \sin (x) + 30 \cos (x)$$

14. Messy, huh? Combine like terms to clean things up a bit:

$$(B - 10A - 31B + 30A) \sin (x) + (-A - 10B + 31A + 30B) \cos (x) = 20 \sin (x) + 30 \cos (x)$$

15. Go ahead and combine terms once more:

$$(-30B + 20A) \sin (x) + (30A + 20B) \cos (x) = 20 \sin (x) + 30 \cos (x)$$

16. Time to start solving! Solving for $A$ gives you

$$A = 1$$

and solving for $B$ gives you

$$B = 0$$

17. So the particular solution is

$$y_p = \sin (x)$$

18. All that's left to do is add the particular solution and the homogeneous solution together to get your general solution of

$$y = c_1 e^{-2x} + c_2 e^{-3x} + c_3 e^{-5x} + \sin (x)$$

**10** **What's the answer to this equation?**

$$y''' + 11y'' + 36y' + 36y = 75 \sin (x) + 105 \cos (x)$$

**Solution: $y = c_1 e^{-2x} + c_2 e^{-3x} + c_3 e^{-6x} + 3 \sin (x)$**

1. Find the homogeneous version of the differential equation:

$$y''' + 11y'' + 36y' + 36y = 0$$

2. Because the homogeneous version has constant coefficients, you can safely assume a solution of this form:

$$y = e^{rx}$$

3. Plugging your attempted solution into the equation gives you

$$r^3 e^{rx} + 11r^2 e^{rx} + 36r e^{rx} + 36 e^{rx} = 0$$

4. Canceling out $e^{rx}$ leaves you with

$$r^3 + 11r^2 + 36r + 36 = 0$$

5. The next step is to factor that cubic equation as follows:

$$(r + 2)(r + 3)(r + 6) = 0$$

6. The roots are

$r_1 = -2$, $r_2 = -3$, and $r_3 = -6$

which means

$y_1 = e^{-2x}$

$y_2 = e^{-3x}$

$y_3 = e^{-6x}$

7. Tada! Your homogeneous solution is

$y_h = c_1 e^{-2x} + c_2 e^{-3x} + c_3 e^{-6x}$

8. You still need to find a particular solution though; go ahead and assume a solution of the form

$y_p = A \sin(x) + B \cos(x)$

9. First, find $y_p'''$:

$y_p''' = -A \cos(x) + B \sin(x)$

10. Then find $11y_p''$:

$11y_p'' = -11A \sin(x) - 11B \cos(x)$

11. Next, find $36y_p'$:

$36y_p' = 36A \cos(x) - 36B \sin(x)$

12. Finally, find $36y_p$:

$36y_p = 36A \sin(x) + 36B \cos(x)$

13. Add together what you found in Steps 9 through 12:

$y''' + 11y'' + 36y' + 36y = -A \cos(x) + B \sin(x) - 11A \sin(x) - 11B \cos(x) + 36A \cos(x) - 36B \sin(x) + 36A \sin(x) + 36B \cos(x) = 75 \sin(x) + 105 \cos(x)$

14. Get rid of that mess by combining terms to get

$(B - 11A - 36B + 36A) \sin(x) + (-A - 11B + 36A + 36B) \cos(x) = 75 \sin(x) + 105 \cos(x)$

15. That equation still isn't ideal to work with, so combine terms further:

$(-35B + 25A) \sin(x) + (35A + 25B) \cos(x) = 75 \sin(x) + 105 \cos(x)$

16. You're almost done. You just need to solve for $A$; doing so gives you

$A = 3$

and for $B$, which gives you

$B = 0$

17. So the particular solution is

$y_p = 3 \sin(x)$

18. For your final act, find the general solution (which is the sum of the homogeneous solution and the particular solution):

$y = c_1 e^{-2x} + c_2 e^{-3x} + c_3 e^{-6x} + 3 \sin(x)$

# Part III

# The Power Stuff: Advanced Techniques

Part III
The Power Stuff:
Advanced Techniques

## In this part . . .

**H**ere's where you improve your ability to use series solutions and Laplace transforms. You also tackle the trick of using differential equation systems to solve problems. Even though there's no one-size-fits-all solution in the differential equations world, if you can use these power techniques and know when to apply them, you'll be set no matter what kind of nasty differential equation gets thrown your way!

# Chapter 8

# Using Power Series to Solve Ordinary Differential Equations

## In This Chapter

▶ Using the ratio test to see whether a series will converge

▶ Practicing your ability to shift the series index

▶ Applying power series to differential equations to find series solutions

*A* power series is an infinite sum of powers of *x*, which you can use to solve differential equations that can't be solved in any other way. In this chapter, you practice working with power series to solve ordinary differential equations. (Keep in mind that this chapter focuses on *ordinary* differential equations; Chapter 9 deals with differential equations with *singular points*. A *singular point* is a value [or values] of *x* where a coefficient in the differential equation goes to infinity. An ordinary differential equation has no singular points.)

Kick things off by practicing the ratio test and shifting the series index. Then put your skills together to solve some equations.

## Checking On a Series with the Ratio Test

Power series that become infinite aren't of much help to anyone, which is why you only work with series that stay finite in the following pages. A *finite* series converges to a particular value.

A series such as the following:

$$y = \sum_{n=0}^{\infty} a_n x^n$$

is said to converge for a particular *x* if this limit:

$$\lim_{n \to \infty} \sum_{n=0}^{m} a_n x^n$$

is finite. If this limit is infinite, the series doesn't converge.

How do you know whether a series converges? Just bust out the *ratio test,* which compares successive terms of a series to see whether the series is going to converge. If the ratio of the $(n + 1)$th term to the $n$th term is less than 1 for a fixed value of $x$, the series converges for that $x$.

For example, if you have this series:

$$y = \sum_{n=0}^{\infty} a_n \left( x - x_0 \right)^n$$

then the ratio of the $(n + 1)$th term to the $n$th term is

$$\frac{a_{n+1} \left( x - x_0 \right)^{n+1}}{a_n \left( x - x_0 \right)^n}$$

The series converges if this ratio is less than 1 as $n$ gets larger and larger.

Here's another example of the ratio test. Take a look and then check out the following problems to practice using the ratio test to determine whether a particular series converges.

**Q.** Does this series converge?

$$\sum_{n=0}^{\infty} \left( -1 \right)^n \left( x - 4 \right)^n$$

**A.** Yes, if $3 < x < 5$.

1. Take a look at the ratio of the $(n + 1)$th term to the $n$th term:

$$\lim_{n \to \infty} \frac{\left| \left( -1 \right)^{n+1} \left( x - 4 \right)^{n+1} \right|}{\left| \left( -1 \right)^n \left( x - 4 \right)^n \right|}$$

2. This ratio becomes

$$\lim_{n \to \infty} \frac{\left| \left( -1 \right)^{n+1} \left( x - 4 \right)^{n+1} \right|}{\left| \left( -1 \right)^n \left( x - 4 \right)^n \right|} = \left| x - 4 \right|$$

3. So the ratio is $|x - 4|$, and the series converges if that ratio is less than 1. In other words, the range in which the series converges is $|x - 4| < 1$.

4. Therefore, if $x$ is in the range $3 < x < 5$, the series converges.

**1.** Does this series converge?

$$\sum_{n=0}^{\infty}(-1)^n(x-5)^n$$

*Solve It*

**2.** State whether this series converges:

$$\sum_{n=0}^{\infty}(-1)^n(x-1)^n$$

*Solve It*

**3.** Does this series converge?

$$\sum_{n=0}^{\infty}(x-2)^{-n}$$

*Solve It*

**4.** State whether this series converges:

$$\sum_{n=0}^{\infty}(x-1)^{-n}$$

*Solve It*

# Shifting the Series Index

Before you can start solving differential equations by using series, you need to be comfortable with a little trick called *shifting the series index,* which allows you to take two series of different indices and make them have the same index so you can compare the two series term by term.

In these two series:

$$y = \sum_{n=0}^{\infty} a_n x^n$$

$$y = \sum_{n=2}^{\infty} a_n (x+1)^n$$

one starts at $n = 0$, and the other starts at $n = 2$, which makes comparing the two series term by term rather difficult. Say you want both series to start at $n = 0$. To do that, just replace $n$ with $n + 2$ in the second series, which gives you

$$y = \sum_{n+2=2}^{\infty} a_{n+2} (x+1)^{n+2}$$

Note that this series index now starts at $n + 2 = 2$. You can subtract 2 from both sides of this expression, which leaves you with

$$y = \sum_{n=0}^{\infty} a_{n+2} (x+1)^{n+2}$$

Tada! The second series now starts at $n = 0$, just like you wanted.

Following are a few practice problems (as well as an additional example) to get you shifting the series index like a math whiz.

*Q.* Shift this series index to start at $n = 0$:

$$\sum_{n=3}^{\infty} (-1)^n (x-4)^n$$

*A.* $\sum_{n=0}^{\infty} (-1)^{n+3} (x-4)^{n+3}$

1. You know that the series starts at $n = 3$, so substitute $n + 3$ for $n$:

$$\sum_{n+3=3}^{\infty} (-1)^{n+3} (x-4)^{n+3}$$

2. Subtract 3 from both sides of the expression to get $n = 0$:

$$\sum_{n=0}^{\infty} (-1)^{n+3} (x-4)^{n+3}$$

**5.** Shift the following series index to start at $n = 0$:

$$\sum_{n=2}^{\infty}(-1)^n(x-1)^n(x-2)^n$$

Solve It

**6.** Make this series start at $n = 0$ by using the shifting the series index technique:

$$\sum_{n=2}^{\infty}(-1)^n(x-2)^{2n}$$

Solve It

**7.** Shift the following series index to start at $n = 0$:

$$\sum_{n=2}^{\infty}\left(4x+9\right)\left(x-2\right)^{4n}$$

*Solve It*

**8.** Make this series start at $n = 0$ by using the shifting the series index technique:

$$\sum_{n=3}^{\infty}\left(x-1\right)^{n}\left(x-2\right)^{2n}\left(x-3\right)^{-n}$$

*Solve It*

# *Exploiting the Power of Power Series to Find Series Solutions*

 Power series are pretty handy for solving ordinary differential equations because you can express just about any solution by using one. When tasked with solving an ordinary differential equation by using a series solution, arm yourself with these essential substitutions:

> ✔ Substitute this series for $y$: $y = \sum_{n=0}^{\infty} a_n x^n$
>
> ✔ Substitute this series for $y'$: $y' = \sum_{n=0}^{\infty} (n+1) a_{n+1} x^n$
>
> ✔ Substitute this series for $y''$: $y'' = \sum_{n=0}^{\infty} (n+2)(n+1) a_{n+2} x^n$

After making these substitutions in your equation, compare the coefficients of $x$ on each side of the equation to solve for the coefficients $a_n$ in the series terms. (And don't forget to use the initial conditions to solve for the coefficients as well.)

Following is an example of this type of problem with each step worked out. I recommend reviewing it before putting your skills to the test solving ordinary differential equations with series solutions.

**Q.** Solve this differential equation by using a series solution:

$$\frac{d^2 y}{dx^2} + y = 0$$

**A.** $y = a_0 \cos(x) + a_1 \sin(x)$

1. Start off with a solution $y$ of the form

$$y = \sum_{n=0}^{\infty} a_n x^n$$

2. To find $y''$, start by finding $y'$. Here's what the terms of the series look like:

$$y = a_0 + a_1 x + a_2 x^2 + a_3 x^3 + \ldots$$

If you differentiate that equation term by term, then $y'$ equals

$$y' = a_1 + 2a_2 x + 3a_3 x^2 + \ldots$$

The general $n$th term here is

$$n a_n x^{n-1}$$

so $y'$ equals

$$y' = \sum_{n=1}^{\infty} n a_n x^{n-1}$$

3. You can find $y''$ by differentiating the $y'$ equation to get

$$y'' = 2a_2 + 6a_3 x + \ldots$$

The general term here is

$$n(n-1) a_n x^{n-2}$$

which means you can state $y''$ as

$$y'' = \sum_{n=2}^{\infty} n(n-1) a_n x^{n-2}$$

Note that this series starts at $n = 2$, not $n = 0$, as the series for $y$ does.

4. Now that you have $y$ and $y''$, go ahead and substitute them into the original differential equation for this result:

$$\sum_{n=2}^{\infty} n(n-1) a_n x^{n-2} + \sum_{n=0}^{\infty} a_n x^n = 0$$

That's your differential equation in series form.

5. To compare these two series, make sure they start at the same index value, $n = 0$. Shift the first series (the one on the left) by replacing $n$ with $n + 2$ to get

$$\sum_{n=0}^{\infty} (n+2)(n+1) a_{n+2} x^n + \sum_{n=0}^{\infty} a_n x^n = 0$$

6. Then combine the two series:

$$\sum_{n=0}^{\infty} \left[ (n+2)(n+1) a_{n+2} x^n + a_n x^n \right] = 0$$

7. Next, factor out $x^n$:

$$\sum_{n=0}^{\infty}\left[(n+2)(n+1)a_{n+2}+a_n\right]x^n = 0$$

8. Because this series equals 0 and must be true for all $x$, each term must equal 0. In other words, you get

$$(n+2)(n+1)\,a_{n+2}+a_n = 0$$

This equation is called a *recurrence relation;* it relates the coefficients of later terms to the coefficients of earlier terms. In particular, you can get all the coefficients in terms of $a_0$ and $a_1$ (which are set by the initial conditions).

9. First, you must determine the even coefficients, which means solving for $a_2$ in terms of $a_0$:

$$(2)(1)a_2 + a_0 = 0$$

so

$$a_2 = \frac{-a_0}{(2)(1)}$$

10. Now find $a_4$:

$$(4)(3)a_4 + a_2 = 0$$

so

$$a_4 = \frac{-a_2}{(4)(3)}$$

11. Because you want the even coefficients in terms of $a_0$, substitute the final equation in Step 9 for $a_2$:

$$a_4 = \frac{a_0}{(4)(3)(2)(1)}$$

Not so fast! Because $(4)(3)(2)(1) = 4!$, you get

$$a_4 = \frac{a_0}{4!}$$

12. For $a_6$, you have

$$(6)(5)a_6 + a_4 = 0$$

or

$$a_6 = \frac{-a_4}{(6)(5)}$$

13. Substituting the final equation from Step 11 for $a_4$ gives you

$$a_6 = \frac{-a_0}{(6)(5)(4!)}$$

But $(6)(5)(4!) = 6!$, so you actually have

$$a_6 = \frac{-a_0}{6!}$$

14. At this point, you know that

$$a_2 = \frac{-a_0}{2!}$$

$$a_4 = \frac{a_0}{4!}$$

$$a_6 = \frac{-a_0}{6!}$$

Believe it or not, you've just generally related the even coefficients! If $n = 2m$ (that is, if $n$ is even), then

$$a_n = a_{2m} = \frac{(-1)^m a_0}{(2m)!} \quad m = 0,1,2,3...$$

15. Now you need to find the odd coefficients. Remember that the recurrence relation for the solution is

$$(n+2)(n+1)a_{n+2}+a_n = 0$$

You can see that for $n = 1$ you get the following:

$$(3)(2)a_3 + a_1 = 0$$

so

$$a_3 = \frac{-a_1}{(3)(2)}$$

16. Similar to what happened with the even coefficients, $(3)(2) = 3!$, so you wind up with

$$a_3 = \frac{-a_1}{3!}$$

17. When you try $n = 3$ in the recurrence relation, you get

$$(5)(4)a_5 + a_3 = 0$$

or

$$a_5 = \frac{-a_3}{(5)(4)}$$

18. Substituting the equation you found in Step 16 for $a_3$ gives you this:

$$a_5 = \frac{a_1}{(5)(4)(3!)}$$

or

$$a_5 = \frac{a_1}{5!}$$

19. When you substitute $n = 5$ into the recurrence relation, you get

$$(7)(6)a_7 + a_5 = 0$$

or

$$a_7 = \frac{-a_5}{(7)(6)}$$

20. Substituting the final equation from Step 18 for $a_5$ leaves you with

$$a_7 = \frac{-a_1}{(7)(6)(5!)}$$

which means that

$$a_7 = \frac{-a_1}{7!}$$

21. To summarize Steps 15 through 20, you now know that

$$a_3 = \frac{-a_1}{3!}$$

$$a_5 = \frac{a_1}{5!}$$

$$a_7 = \frac{-a_1}{7!}$$

22. If $n = 2m + 1$, you can generally relate the odd coefficients as follows:

$$a_n = a_{2m+1} = \frac{(-1)^m a_1}{(2m+1)!} \quad m = 0,1,2,3\ldots$$

23. You can now write the whole solution as

$$y = a_0 \sum_{m=0}^{\infty} \frac{(-1)^m x^{2m}}{(2m)!} + a_1 \sum_{m=0}^{\infty} \frac{(-1)^m x^{2m+1}}{(2m+1)!}$$

24. Surprise! The two series are recognizable as $\cos(x)$ and $\sin(x)$:

$$\sum_{n=0}^{\infty} \frac{(-1)^n x^{2n}}{(2n)!} = \cos(x)$$

and

$$\sum_{n=0}^{\infty} \frac{(-1)^n x^{2n+1}}{(2n+1)!} = \sin(x)$$

25. After all that work, you can write the solution as

$$y = a_0 \cos(x) + a_1 \sin(x)$$

**9.** Solve this differential equation by using a series solution:

$$\frac{d^2y}{dx^2} + 4y = 0$$

Solve It

**10.** Find the solution to this differential equation by using a series solution:

$$\frac{d^2y}{dx^2} - y = 0$$

Solve It

# Answers to Solving Ordinary Differential Equations with Power Series

Here are the answers to the practice questions I provide throughout this chapter. I walk you through each answer so you can see the problems worked out step by step. Enjoy!

**1** **Does this series converge?**

$$\sum_{n=0}^{\infty}(-1)^n(x-5)^n$$

**Answer: Yes, if 4 < x < 6.**

1. Take a look at the ratio of the (n + 1)th term to the nth term:

$$\lim_{n\to\infty}\frac{\left|(-1)^{n+1}(x-5)^{n+1}\right|}{\left|(-1)^n(x-5)^n\right|}$$

2. This ratio becomes

$$\lim_{x\to\infty}\frac{\left|(-1)^{n+1}(x-5)^{n+1}\right|}{\left|(-1)^n(x-5)^n\right|}=|x-5|$$

3. So the ratio is |x − 5|, and the series converges if that ratio is less than 1. In other words, the range in which the series converges is |x − 5| < 1.

4. Therefore, if x is in the range 4 < x < 6, the series converges.

**2** **State whether this series converges:**

$$\sum_{n=0}^{\infty}(-1)^n(x-1)^n$$

**Answer: Yes, if 0 < x < 2.**

1. Check out the ratio of the (n + 1)th term to the nth term:

$$\lim_{x\to\infty}\frac{\left|(-1)^{n+1}(x-1)^{n+1}\right|}{\left|(-1)^n(x-1)^n\right|}$$

2. This ratio works out to

$$\lim_{x\to\infty}\frac{\left|(-1)^{n+1}(x-1)^{n+1}\right|}{\left|(-1)^n(x-1)^n\right|}=|x-1|$$

3. As you can see, the ratio is |x − 1|; the series converges if that ratio is less than 1. So the range in which the series converges absolutely is |x − 1| < 1.

4. Thus, if x is in the range 0 < x < 2, the series converges.

**3** **Does this series converge?**

$$\sum_{n=0}^{\infty}(x-2)^{-n}$$

**Answer: Yes, if $x > 3$ or $x < 1$.**

1. Take a look at the ratio of the $(n + 1)$th term to the $n$th term:

$$\lim_{x \to \infty} \frac{\left|(x-2)^n\right|}{\left|(x-2)^{n+1}\right|}$$

2. This ratio becomes

$$\lim_{x \to \infty} \frac{\left|(x-2)^n\right|}{\left|(x-2)^{n+1}\right|} = \left|x-2\right|^{-1}$$

3. So the ratio is $|x - 2|^{-1}$, and the series converges if that ratio is less than 1. In other words, the range in which the series converges is $|x - 2|^{-1} < 1$.

4. Therefore, if $x$ is in the range $x > 3$ or $x < 1$, the series converges.

**4** **State whether this series converges:**

$$\sum_{n=0}^{\infty}(x-1)^{-n}$$

**Answer: Yes, if $x > 2$ or $x < 0$.**

1. Check out the ratio of the $(n + 1)$th term to the $n$th term:

$$\lim_{x \to \infty} \frac{\left|(x-1)^n\right|}{\left|(x-1)^{n+1}\right|}$$

2. This ratio works out to

$$\lim_{x \to \infty} \frac{\left|(x-1)^n\right|}{\left|(x-1)^{n+1}\right|} = \left|x-1\right|^{-1}$$

3. As you can see, the ratio is $|x - 1|^{-1}$; the series converges if that ratio is less than 1. So the range in which the series converges absolutely is $|x - 1|^{-1} < 1$.

4. Thus, if $x$ is in the range $x > 2$ or $x < 0$, the series converges.

**5** **Shift the following series index to start at $n = 0$:**

$$\sum_{n=2}^{\infty}(-1)^n(x-1)^n(x-2)^n$$

**Solution:** $\sum_{n=0}^{\infty}(-1)^{n+2}(x-1)^{n+2}(x-2)^{n+2}$

1. The series starts at $n = 2$, so substitute $n + 2$ for $n$:

$$\sum_{n+2=2}^{\infty}(-1)^{n+2}(x-1)^{n+2}(x-2)^{n+2}$$

2. Subtract 2 from both sides of the expression to get $n = 0$:

$$\sum_{n=0}^{\infty}(-1)^{n+2}(x-1)^{n+2}(x-2)^{n+2}$$

That's it! Your work here (on this problem anyway) is done.

**6** **Make this series start at $n = 0$ by using the shifting the series index technique:**

$$\sum_{n=2}^{\infty}(-1)^n(x-2)^{2n}$$

**Solution:** $\sum_{n=0}^{\infty}(-1)^{n+2}(x-2)^{2n+4}$

1. You know that the series begins at $n = 2$. That fact is your clue to substitute $n + 2$ for $n$:

$$\sum_{n+2=2}^{\infty}(-1)^{n+2}(x-2)^{2(n+2)}$$

2. To set $n$ equal to 0, simply subtract 2 from both sides of the expression:

$$\sum_{n=0}^{\infty}(-1)^{n+2}(x-2)^{2n+4}$$

**7** **Shift the following series index to start at $n = 0$:**

$$\sum_{n=2}^{\infty}(4x+9)(x-2)^{4n}$$

**Solution:** $\sum_{n=0}^{\infty}(4x+9)(x-2)^{4n+8}$

1. The series starts at $n = 2$, so substitute $n + 2$ for $n$:

$$\sum_{n+2=2}^{\infty}(4x+9)(x-2)^{4(n+2)}$$

2. Subtract 2 from both sides of the expression to get $n = 0$:

$$\sum_{n=0}^{\infty}(4x+9)(x-2)^{4n+8}$$

**8** **Make this series start at $n = 0$ by using the shifting the series index technique:**

$$\sum_{n=3}^{\infty}(x-1)^{n}(x-2)^{2n}(x-3)^{-n}$$

**Solution:** $\sum_{n=0}^{\infty}(x-1)^{n+3}(x-2)^{2(n+3)}(x-3)^{-(n+3)}$

1. You know that the series begins at $n = 3$. That fact is your clue to substitute $n + 3$ for $n$:

$$\sum_{n+3=3}^{\infty}(x-1)^{n+3}(x-2)^{2(n+3)}(x-3)^{-(n+3)}$$

2. To set $n$ equal to 0, simply subtract 3 from both sides of the expression:

$$\sum_{n=0}^{\infty}(x-1)^{n+3}(x-2)^{2(n+3)}(x-3)^{-(n+3)}$$

**9** **Solve this differential equation by using a series solution:**

$$\frac{d^2y}{dx^2} + 4y = 0$$

**Solution:** $y = a_0 \cos (2x) + a_1 \sin (2x)$

1. Start off with a solution $y$ of the form

$$y = \sum_{n=0}^{\infty}a_n x^n$$

2. To find $y''$, start by finding $y'$. Here's what the terms of the series look like:

$$y = a_0 + a_1 x + a_2 x^2 + a_3 x^3 + \ldots$$

If you differentiate that equation term by term, then $y'$ equals

$$y' = a_1 + 2a_2 x + 3a_3 x^2 + \ldots$$

The general $n$th term here is

$$na_n x^{n-1}$$

so $y'$ equals

$$y' = \sum_{n=1}^{\infty}na_n x^{n-1}$$

3. You can find $y''$ by differentiating the $y'$ equation to get

$$y'' = 2a_2 + 6a_3x + \ldots$$

The general term here is

$$n(n-1)a_nx^{n-2}$$

which means you can state $y''$ as

$$y'' = \sum_{n=2}^{\infty} n(n-1)a_nx^{n-2}$$

Note that this series starts at $n = 2$, not $n = 0$, as the series for $y$ does.

4. Now that you have $y$ and $y''$, go ahead and substitute them into the original differential equation for this result:

$$\sum_{n=2}^{\infty} n(n-1)a_nx^{n-2} + 4\sum_{n=0}^{\infty} a_nx^n = 0$$

That's your differential equation in series form.

5. To compare these two series, make sure they start at the same index value, $n = 0$. Shift the first series (the one on the left) by replacing $n$ with $n + 2$ to get

$$\sum_{n=0}^{\infty} (n+2)(n+1)a_{n+2}x^n + 4\sum_{n=0}^{\infty} a_nx^n = 0$$

6. Then combine the two series:

$$\sum_{n=0}^{\infty} \left[ (n+2)(n+1)a_{n+2}x^n + 4a_nx^n \right] = 0$$

7. Next, factor out $x_n$:

$$\sum_{n=0}^{\infty} \left[ (n+2)(n+1)a_{n+2} + 4a_n \right] x^n = 0$$

8. Because this series equals 0 and must be true for all $x$, each term must equal 0. In other words, you get

$$(n + 2)(n + 1)a_{n+2} + 4a_n = 0$$

This equation is called a *recurrence relation;* it relates the coefficients of later terms to the coefficients of earlier terms. In particular, you can get all the coefficients in terms of $a_0$ and $a_1$ (which are set by the initial conditions).

9. Your first step in relating coefficients is finding all the even coefficients in terms of $a_0$. Start by solving for $a_2$ in terms of $a_0$:

$$(2)(1)a_2 + 4a_0 = 0$$

so

$$a_2 = \frac{-4a_0}{(2)(1)}$$

10. Now find $a_4$:

$$(4)(3)a_4 + 4a_2 = 0$$

so

$$a_4 = \frac{-4a_2}{(4)(3)}$$

11. Because you want the even coefficients in terms of $a_0$, substitute the final equation in Step 9 for $a_2$:

$$a_4 = \frac{4a_0}{(4)(3)(2)(1)}$$

Not so fast! Because $(4)(3)(2)(1) = 4!$, you get

$$a_4 = \frac{4a_0}{4!}$$

12. For $a_6$, you have

$$(6)(5)a_6 + 4a_4 = 0$$

or

$$a_6 = \frac{-4a_4}{(6)(5)}$$

13. Substituting the final equation in Step 11 for $a_4$ gives you

$$a_6 = \frac{-4a_0}{(6)(5)(4!)}$$

But $(6)(5)(4!) = 6!$, so you actually have

$$a_6 = \frac{-4a_0}{6!}$$

14. At this point, you know that

$$a_2 = \frac{-4a_0}{2!}$$

$$a_4 = \frac{4a_0}{4!}$$

$$a_6 = \frac{-4a_0}{6!}$$

Believe it or not, you've just generally related the even coefficients! If $n = 2m$ (that is, if $n$ is even), then

$$a_n = a_{2m} = \frac{4(-1)^m a_0}{(2m)!} \quad m = 0,1,2,3\ldots$$

15. Now you need to find the odd coefficients. Remember that the recurrence relation for the solution is

$$(n + 2)(n + 1)a_{n+2} + 4a_n = 0$$

You can see that for $n = 1$, you get the following:

$$(3)(2)a_3 + 4a_1 = 0$$

so

$$a_3 = \frac{-4a_1}{(3)(2)}$$

16. Similar to what happened with the even coefficients, $(3)(2) = 3!$, so you wind up with

$$a_3 = \frac{-4a_1}{3!}$$

17. When you try $n = 3$ in the recurrence relation, you get

$$(5)(4)a_5 + 4a_3 = 0$$

or

$$a_5 = \frac{-4a_3}{(5)(4)}$$

18. Substituting the equation you found in Step 16 for $a_3$ gives you this:

$$a_5 = \frac{4a_1}{(5)(4)(3!)}$$

or

$$a_5 = \frac{4a_1}{5!}$$

19. When you substitute $n = 5$ into the recurrence relation, you get

$$(7)(6)a_7 + 4a_5 = 0$$

or

$$a_7 = \frac{-4a_5}{(7)(6)}$$

20. Substituting the final equation from Step 18 for $a_5$ leaves you with

$$a_7 = \frac{-4a_1}{(7)(6)(5!)}$$

which means that

$$a_7 = \frac{-4a_1}{7!}$$

21. To summarize Steps 15 through 20, you now know that

$$a_3 = \frac{-4a_1}{3!}$$

$$a_5 = \frac{4a_1}{5!}$$

$$a_7 = \frac{-4a_1}{7!}$$

22. If $n = 2m + 1$, you can generally relate the odd coefficients as follows:

$$a_n = a_{2m+1} = \frac{4(-1)^m a_1}{(2m+1)!} \quad m = 0,1,2,3...$$

23. You can now write the whole solution as

$$y = a_0 \sum_{m=0}^{\infty} \frac{4(-1)^m x^{2m}}{(2m)!} + a_1 \sum_{m=0}^{\infty} \frac{4(-1)^m x^{2m+1}}{(2m+1)!}$$

24. Move the factor of 4 to the $x$ term:

$$y = a_0 \sum_{m=0}^{\infty} \frac{(-1)^m (2x)^{2m}}{(2m)!} + a_1 \sum_{m=0}^{\infty} \frac{(-1)^m (2x)^{2m+1}}{(2m+1)!}$$

25. In this case, the two series are recognizable as $\cos(2x)$ and $\sin(2x)$:

$$\sum_{n=0}^{\infty} \frac{(-1)^n (2x)^{2n}}{(2n)!} = \cos(2x)$$

and

$$\sum_{n=0}^{\infty} \frac{(-1)^n (2x)^{2n+1}}{(2n+1)!} = \sin(2x)$$

26. After all that work, you can write the solution as

$$y = a_0 \cos(2x) + a_1 \sin(2x)$$

**10** **Find the solution to this differential equation by using a series solution:**

$$\frac{d^2y}{dx^2} - y = 0$$

**Solution: $y = c_0 e^x + c_1 e^{-x}$**

1. Begin with a solution $y$ of the following form:

$$y = \sum_{n=0}^{\infty} a_n x^n$$

2. Your next step is to use $y'$ to find $y''$. Here's what the terms of the series look like:

$$y = a_0 + a_1 x + a_2 x^2 + a_3 x^3 + \ldots$$

When you differentiate term by term, $y'$ equals

$$y' = a_1 + 2a_2 x + 3a_3 x^2 + \ldots$$

The general $n$th term here is

$$na_n x^{n-1}$$

which makes $y'$ equal to

$$y' = \sum_{n=1}^{\infty} na_n x^{n-1}$$

3. Find $y''$ by differentiating the $y'$ equation:

$$y'' = 2a_2 + 6a_3 x + \ldots$$

The general term here is

$$n(n-1)a_n x^{n-2}$$

which means you can write $y''$ as

$$y'' = \sum_{n=2}^{\infty} n(n-1)a_n x^{n-2}$$

Note that this series starts at $n = 2$, not $n = 0$, as the series for $y$ does.

4. Refer to the original differential equation and substitute in $y$ and $y''$ to get the differential equation in series form:

$$\sum_{n=2}^{\infty} n(n-1)a_n x^{n-2} - \sum_{n=0}^{\infty} a_n x^n = 0$$

5. To compare these series, make sure they start at the same index value, $n = 0$. You can do so by shifting the first series index; just replace $n$ with $n + 2$. The result is

$$\sum_{n=0}^{\infty} (n+2)(n+1)a_{n+2} x^n - \sum_{n=0}^{\infty} a_n x^n = 0$$

6. Next up, combine the two series to get

$$\sum_{n=0}^{\infty} \left[ (n+2)(n+1)a_{n+2} x^n - a_n x^n \right] = 0$$

7. Then factor out $x^n$:

$$\sum_{n=0}^{\infty}\left[(n+2)(n+1)a_{n+2}-a_n\right]x^n = 0$$

8. This series equals 0 and must be true for all $x$, so each term must equal 0. In other words, you wind up with

$$(n+2)(n+1)a_{n+2}-a_n = 0$$

which is called a *recurrence relation;* it relates the coefficients of later terms to the coefficients of earlier terms. You can actually use this relation to get all the coefficients in terms of $a_0$ and $a_1$ (which are set by the initial conditions).

9. Your first step in relating coefficients is finding all the even coefficients in terms of $a_0$. Start by solving for $a_2$:

$$(2)(1)a_2 - a_0 = 0$$

or

$$a_2 = \frac{a_0}{(2)(1)}$$

10. Great. Now find $a_4$:

$$(4)(3)a_4 - a_2 = 0$$

so

$$a_4 = \frac{a_2}{(4)(3)}$$

11. Substituting the final equation in Step 9 for $a_2$ gives you

$$a_4 = \frac{a_0}{(4)(3)(2)(1)}$$

Ah, but $(4)(3)(2)(1) = 4!$, so in reality you actually have

$$a_4 = \frac{a_0}{4!}$$

12. Now for $a_6$:

$$(6)(5)a_6 - a_4 = 0$$

so

$$a_6 = \frac{a_4}{(6)(5)}$$

13. When you substitute the final equation from Step 11 for $a_4$, you get

$$a_6 = \frac{a_0}{(6)(5)(4!)}$$

Wait a second. Because $(6)(5)(4!) = 6!$, you really have

$$a_6 = \frac{a_0}{6!}$$

14. To sum up Steps 9 through 13, you know that

$$a_2 = \frac{a_0}{2!}$$

$$a_4 = \frac{a_0}{4!}$$

$$a_6 = \frac{a_0}{6!}$$

All of that together generally relates the even coefficients:

$$a_n = \frac{a_0}{n!} \quad n = 0, 2, 4, 6\ldots$$

15. Time to turn to the odd coefficients. Because the recurrence relation for the solution is

$$(n + 2)(n + 1)a_{n+2} - a_n = 0$$

you can see that for $n = 1$ you get

$$(3)(2)a_3 - a_1 = 0$$

so

$$a_3 = \frac{a_1}{(3)(2)}$$

16. No surprise that $(3)(2) = 3!$, right? So you're left with

$$a_3 = \frac{a_1}{3!}$$

17. Plug $n = 3$ into the recurrence relation to get

$$(5)(4)a_5 - a_3 = 0$$

or

$$a_5 = \frac{a_3}{(5)(4)}$$

18. When you substitute the equation from Step 16 for $a_3$, you wind up with

$$a_5 = \frac{a_1}{(5)(4)(3!)}$$

so

$$a_5 = \frac{a_1}{5!}$$

19. Substituting $n = 5$ into the recurrence relation gives you

$$(7)(6)a_7 - a_5 = 0$$

or

$$a_7 = \frac{a_5}{(7)(6)}$$

20. Substituting the final equation from Step 18 for $a_5$ leaves you with

$$a_7 = \frac{a_1}{(7)(6)(5!)}$$

which means

$$a_7 = \frac{a_1}{7!}$$

21. At this point, you know that

$$a_3 = \frac{a_1}{3!}$$

$$a_5 = \frac{a_1}{5!}$$

$$a_7 = \frac{a_1}{7!}$$

22. You can now generally write the odd coefficients of the solution as follows, if $n = 2m + 1$:

$$a_n \frac{a_1}{(n)!} \quad m = 1,3,5,7...$$

23. So the whole solution can actually be written as

$$y = a_0 \sum_{m=0}^{\infty} \frac{x^{2m}}{(2m)!} + a_1 \sum_{m=0}^{\infty} \frac{x^{2m+1}}{(2m+1)!}$$

Technically that's your answer, but you can keep going so as to get the solution into an easier form.

24. In fact, the series are equal to

$$\cosh(x) = \sum_{m=0}^{\infty} \frac{x^{2m}}{(2m)!}$$

and

$$\sinh(x) = \sum_{m=0}^{\infty} \frac{x^{2m+1}}{(2m+1)!}$$

which means you can write the solution as

$$y = a_0 \cosh(x) + a_1 \sinh(x)$$

25. Write $\sinh(x)$ and $\cosh(x)$ in terms of exponentials as follows:

$$\sinh(x) = \frac{e^x - e^{-x}}{2}$$

and

$$\cosh(x) = \frac{e^x - e^{-x}}{2}$$

26. Because $\sinh(x)$ and $\cosh(x)$ can be written as a sum of exponentials, you can rewrite the solution as

$$y = c_0 e^x + c_1 e^{-x}$$

where $c_0$ and $c_1$ are determined by the initial conditions. Nice work!

# Chapter 9

# Solving Differential Equations with Series Solutions Near Singular Points

. . . . . . . . . . . . . . . . . . . . . . . . . . . . . . . . . . . . . . . . . . . . . . . . . . . . . . . . . . . . . . . .

### In This Chapter
▶ Identifying singular points
▶ Determining whether singular points are regular or irregular
▶ Getting familiar with Euler's equation
▶ Solving general differential equations that look like Euler equations

. . . . . . . . . . . . . . . . . . . . . . . . . . . . . . . . . . . . . . . . . . . . . . . . . . . . . . . . . . . . . . . .

**D**ifferential equations can get a little unruly when terms within them head off into infinity (meaning they become *unbounded*). The points where functions go to infinity are called *singular points,* and you can expect to encounter both regular and irregular singular points in your dabbling with differential equations.

The good news is you can handle regular singular points by using the methods in this chapter. As for irregular singular points? Forget 'em. They're a lost cause because their differential equations can't be solved near such irregular singular points.

This chapter has you find singular points, classify them as regular or irregular, and go about solving them. And to make solving differential equations near regular singular points that much easier, this chapter also shows you how to deal with a special class of differential equations called Euler equations. You can often create a series expansion around a known solution to Euler's equation for a general differential equation with regular singular points.

## Finding Singular Points

*Singular points* occur when a coefficient in a particular differential equation becomes unbounded.

For example, in this differential equation

$$\frac{d^2 y}{dx^2} + p(x)\frac{dy}{dx} + q(x)y = 0$$

where

$$p(x) = Q(x)/P(x)$$

and

$$q(x) = R(x)/P(x)$$

the singular points occur where $Q(x)/P(x)$ and/or $R(x)/P(x)$ become unbounded.

In the following problems, you practice finding singular points in differential equations. But first, a quick example.

**Q.** What are the singular points of this differential equation?

$$(4 - x^2)\frac{d^2 y}{dx^2} + x^3\frac{dy}{dx} + (1 + x)y = 0$$

**A.** $x_1 = 2$ and $x_2 = -2$

1. First, put the equation into the following form:

$$\frac{d^2 y}{dx^2} + p(x)\frac{dy}{dx} + q(x)y = 0$$

where

$$p(x) = Q(x)/P(x)$$

and

$$q(x) = R(x)/P(x)$$

Doing so gives you

$$\frac{d^2 y}{dx^2} + \frac{x^3}{(4 - x^2)}\frac{dy}{dx} + \frac{(1 + x)y}{(4 - x^2)} = 0$$

2. Therefore

$$p(x) = \frac{x^3}{(4 - x^2)}$$

and

$$q(x) = \frac{(1 + x)}{(4 - x^2)}$$

3. Looks like $p(x)$ and $q(x)$ both become unbounded when $4 - x^2 = 0$, so the singular points are

$$x_1 = 2 \text{ and } x_2 = -2$$

**1.** What are the singular points of this differential equation?

$$x^2 \frac{d^2y}{dx^2} + \left(8 - x^3\right) \frac{dy}{dx} + \left(1 - 9x\right)y = 0$$

*Solve It*

**2.** Solve for the singular points of this equation:

$$\left(x - 9\right) \frac{d^2y}{dx^2} - x^3 \frac{dy}{dx} + \frac{1}{x^2}y = 0$$

*Solve It*

**3.** What are the singular points of this differential equation?

$$x^2 \frac{d^2 y}{dx^2} + x^3 \frac{dy}{dx} + x^2 y = 0$$

Solve It

**4.** Solve for the singular points of this equation:

$$\left(x^2 + 3x + 2\right) \frac{d^2 y}{dx^2} + \frac{dy}{dx} + y = 0$$

Solve It

# *Classifying Singular Points as Regular or Irregular*

Singular points come in two different forms: regular and irregular. *Regular singular points* are well-behaved and defined in terms of the ratio $Q(x)/P(x)$ and $R(x)/P(x)$, where $P(x)$, $Q(x)$, and $R(x)$ are the polynomial coefficients in the differential equation you're trying to solve.

Irregular singular points are a totally different ball game — and one that I don't get into in this chapter. As you work through the practice problems here, if the singular point in question doesn't appear to be regular, you know it's irregular.

Allow me to introduce you to this dainty differential equation:

$$P(x)\,\frac{d^2y}{dx^2}+Q(x)\,\frac{dy}{dx}+R(x)y=0$$

In order for $x_0$ to be a regular singular point, these two relations must be true:

$$\lim_{x \to x_0}\left(x-x_0\right)\frac{Q(x)}{P(x)} \quad \text{remains finite}$$

and

$$\lim_{x \to x_0}\left(x-x_0\right)^2\frac{R(x)}{P(x)} \quad \text{remains finite}$$

If you define

$$p(x) = Q(x)/P(x)$$

and

$$q(x) = R(x)/P(x)$$

then the two limits become

$$\lim_{x \to x_0}\left(x-x_0\right)p(x) \quad \text{remains finite}$$

and

$$\lim_{x \to x_0}\left(x-x_0\right)^2 q(x) \quad \text{remains finite}$$

If both of these statements are true, then the point $x_0$ is a regular singular point.

In the following problems, you practice classifying singular points as regular or irregular. Don't worry — it's actually kind of fun!

**Q.** Are the singular points of this differential equation regular or irregular?

$$2\frac{d^2y}{dx^2} + \frac{4}{x^2}y = 0$$

**A.** The singular points are regular.

1. Start solving by putting the differential equation into the form

$$\frac{d^2y}{dx^2} + p(x)\frac{dy}{dx} + q(x)y = 0$$

where

$$p(x) = Q(x)/P(x)$$

and

$$q(x) = R(x)/P(x)$$

2. Now you have

$$\frac{d^2y}{dx^2} + \frac{2}{x^2}y = 0$$

which means

$$p(x) = 0$$

and

$$q(x) = \frac{2}{x^2}$$

So the singular points are $x_1 = 0$ and $x_2 = 0$.

3. Evaluate:

$$\lim_{x \to x_0}(x - x_0)p(x)$$

The singular points, $x_0$, are both 0, so this equation becomes

$$\lim_{x \to 0} 0$$

and the value of the expression is 0, which is finite.

4. Next, evaluate the following:

$$\lim_{x \to x_0}(x - x_0)^2 q(x)$$

Again, the singular points, $x_0$, are both 0, so this equation becomes

$$\lim_{x \to 0} 2$$

and the value of this expression is 2, which is finite.

5. Because the limits are finite, the singular points are regular. That wasn't too hard, was it?

**5.** Are the singular points of this differential equation regular or irregular?

$$x \frac{d^2y}{dx^2} + \frac{y}{x} = 0$$

*Solve It*

**6.** Determine whether the singular points of this equation are regular or irregular:

$$\frac{d^2y}{dx^2} + \frac{y}{(x^2 - 4)} = 0$$

*Solve It*

# Working with Euler's Equation

Solving Euler's equation allows you to find the solution to many differential equations with regular singular points. If you can put a differential equation into a similar form to Euler's equation, you can use a series expansion around the solutions you develop for Euler's equation. That's a pretty cool trick in my book.

When you want to play nice with Euler's equation, start by assuming a solution of this form: $y = x^r$.

Then substitute that solution into Euler's equation:

$$[r(r-1) + \alpha r + \beta]\, x^r = 0$$

$$r(r-1) + \alpha r + \beta = 0$$

$$r^2 - r + \alpha r + \beta = 0$$

Ultimately, you wind up with

$$r^2 + (\alpha - 1)\, r + \beta = 0$$

The roots, $r_1$ and $r_2$, of this equation are

$$r_1, r_2 = \frac{-(\alpha - 1) \pm \sqrt{(\alpha - 1)^2 - 4\beta}}{2}$$

You don't know what $\alpha$ and $\beta$ are, so you're forced to consider three cases for these roots:

- $r_1$ and $r_2$ are real and distinct.
- $r_1$ and $r_2$ are real and equal.
- $r_1$ and $r_2$ are complex conjugates.

Following are examples to illustrate each of the three cases. When you're done reviewing them, get more acquainted with Euler's equation in the practice problems.

**Q.** Solve this differential equation:

$$6x^2 \frac{d^2y}{dx^2} + 9x \frac{dy}{dx} - 3y = 0$$

**A.** $y = c_1 x^{1/2} + c_2 x^{-1}$

1. This differential equation has the form of Euler's equation:

   $$x^2 \frac{d^2y}{dx^2} + \alpha x \frac{dy}{dx} + \beta y = 0$$

   which means you can assume its solution looks like

   $$y = x^r$$

2. Substituting your attempted solution into the differential equation gives you

   $$[6r(r-1) + 9r - 3] x^r = 0$$

   or

   $$[6r(r-1) + 9r - 3] = 0$$

   which is actually

   $$6r^2 + 9r - 3 = 0$$

3. Factor this equation to get

   $$3(2r-1)(r+1) = 0$$

4. It appears that the roots are

   $$r_1 = {}^1/_2 \text{ and } r_2 = -1$$

5. So the general solution to the original equation is

   $$y = c_1 x^{1/2} + c_2 x^{-1}$$

**Q.** Find the solution to this differential equation:

$$x^2 \frac{d^2y}{dx^2} + 5x \frac{dy}{dx} + 4y = 0$$

**A.** $y = c_1 x^{-2} + c_2 \ln(x) x^{-2}$

1. Notice that the equation in the question has the form of Euler's equation:

   $$x^2 \frac{d^2y}{dx^2} + \alpha x \frac{dy}{dx} + \beta y = 0$$

2. Go ahead and assume that the solution is of this form:

   $$y = x^r$$

3. Then substitute your attempted solution into the equation:

   $$[r(r-1) + 5r + 4] x^r = 0$$

   or

   $$[r(r-1) + 5r + 4] = 0$$

   which is

   $$r^2 + 4r - 4 = 0$$

4. Factoring this equation gives you

   $$(r+2)(r+2) = 0$$

5. So the roots are

   $$r_1 = -2 \text{ and } r_2 = -2$$

6. Therefore, the general solution to the differential equation is

   $$y = c_1 x^{-2} + c_2 \ln(x) x^{-2}$$

**Q.** Solve this differential equation:

$$x^2 \frac{d^2y}{dx^2} + x \frac{dy}{dx} + y = 0$$

**A.** $y = c_1 \cos (\ln |x|) + c_2 \sin (\ln |x|), x \neq 0$

1. I bet you can guess by now that this differential equation has the form of Euler's equation:

$$x^2 \frac{d^2y}{dx^2} + \alpha x \frac{dy}{dx} + \beta y = 0$$

2. Next up, assume that the solution is of the form

$$y = x^r$$

3. Substituting your attempted solution into the equation gives you

$$[r(r-1) + r + 1] x^r = 0$$

or

$$[r(r-1) + r + 1] = 0$$

which is actually

$$r^2 + 1 = 0$$

4. So the roots are

$$r_1 = i \text{ and } r_2 = -i$$

5. All that mumbo jumbo means the general solution to the differential equation is

$$y = c_1 \cos (\ln |x|) + c_2 \sin (\ln |x|), x \neq 0$$

**7.** Find the solution to this differential equation:

$$x^2 \frac{d^2y}{dx^2} + 4x \frac{dy}{dx} + 2y = 0$$

*Solve It*

**8.** Solve this differential equation:

$$x^2 \frac{d^2y}{dx^2} + 5x \frac{dy}{dx} + 3y = 0$$

*Solve It*

**9.** Find the solution to this differential equation:

$$x^2 \frac{d^2 y}{dx^2} + 6x \frac{dy}{dx} + 6y = 0$$

*Solve It*

**10.** Solve this differential equation:

$$x^2 \frac{d^2 y}{dx^2} + 3x \frac{dy}{dx} + y = 0$$

*Solve It*

# Solving General Differential Equations with Regular Singular Points

This section is where you get to put together everything you practice throughout this chapter (so if you haven't reviewed the previous sections, you may want to flip back a few pages). All set? Then take a look at this general differential equation and assume that it has a regular singular point at $x = 0$:

$$\frac{d^2 y}{dx^2} - p(x)\frac{dy}{dx} + q(x)y = 0$$

Multiply by $x^2$ to make sure that the terms $xp(x)$ and $x^2 q(x)$ (at least one of which has a singular point at $x = 0$) can be expanded into a series:

$$xp(x) = \sum_{n=0}^{\infty} p_n x^n$$

and

$$x^2 q(x) = \sum_{n=0}^{\infty} q_n x^n$$

The Euler equation that matches the general differential equation you're handling is

$$x^2 \frac{d^2 y}{dx^2} - p_0 x \frac{dy}{dx} + q_0 y = 0$$

which means you can assume the Euler equation has a solution of the form

$$y = x^r$$

To handle the fact that the differential equation you're working with isn't an exact Euler equation, add a series expansion to the solution (with the assumption that the $n = 0$ term is the largest term and that all other terms diminish rapidly):

$$y = \sum_{n=0}^{\infty} q_n x^{n+r}$$

The result is the form of your anticipated solution, based on the assumption that the differential equation you're trying to solve isn't too different from Euler's equation.

Substituting $y$, $y'$, and $y''$ into the differential equation gives you

$$a_0 f(r)x^r + \sum_{n=1}^{\infty}\left[ f(n+r)a_n + \sum_{m=0}^{n-1} a_m \left[ (m+r)p_{n-m} + q_{n-m}\right]\right]x^{n+r} = 0$$

where

$$f(r) = r(r+1) + p_0 r + q_0$$

and

$$p_0 = \lim_{x\to 0} xp(x)$$

and

$$q_0 = \lim_{x\to 0} x^2 q(x)$$

Putting all this information together gives you the following recurrence relation (see Chapter 8 for more practice with recurrence relations), which allows you to find the coefficients $a_n$:

$$a_n = \frac{-\sum_{m=0}^{n-1} a_m\left[(m+r)p_{n-m} + q_{n-m}\right]}{f(n+r)} \quad n \geq 1$$

*Note:* If you're not sure how to put everything together to achieve the preceding equation, check out *Differential Equations For Dummies* (Wiley) for the formal theorem.

The differential equation you're solving for has two solutions, $y_1$ and $y_2$, which correspond to the two roots, $r_1$ and $r_2$, of its characteristic equation. The first solution, $y_1$, is

$$y_1 = |x^{r_1}|\left[ 1 + \sum_{n=1}^{\infty} a_n(r_1)x^n\right] \quad x \neq 0$$

where $a_n(r_1)$ are the coefficients using the first root, $r_1$.

If $r_2 \neq r_1$, and if $r_1$ and $r_2$ don't differ by an integer, then the second solution is given by

$$y_2 = |x^{r_2}|\left[ 1 + \sum_{n=1}^{\infty} a_n(r_2)x^n\right] \quad x \neq 0$$

where $a_n(r_2)$ are the coefficients using the second root, $r_2$.

The following problems give you practice solving differential equations that look very much like Euler's equation.

**Q.** Determine the Euler equation that this differential equation is similar to; then determine the Euler equation's two roots, $r_1$ and $r_2$:

$$2x^2 \frac{d^2y}{dx^2} - x \frac{dy}{dx} + (1+x)y = 0$$

**A.** $2x^2 \frac{d^2y}{dx^2} - x \frac{dy}{dx} + y = 0$   roots = 1, $\frac{1}{2}$

1. First, put the differential equation into this form:

$$x^2 \frac{d^2y}{dx^2} - x\left[ xp(x) \right] \frac{dy}{dx} + \left[ x^2 q(x) \right] y = 0$$

Doing so gives you

$$p(x) = \frac{1}{x}$$

and

$$q(x) = \frac{(1+x)}{x^2}$$

2. The Euler equation most similar to your differential equation is

$$x^2 \frac{d^2y}{dx^2} - p_0 x \frac{dy}{dx} + q_0 y = 0$$

where

$$p_0 = \lim_{x \to 0} xp(x)$$

and

$$q_0 = \lim_{x \to 0} x^2 q(x)$$

so

$$p_0 = 1 \text{ and } q_0 = 1$$

3. Therefore, the Euler equation is

$$2x^2 \frac{d^2y}{dx^2} - x \frac{dy}{dx} + y = 0$$

4. To find the roots of the Euler equation, substitute this form of $y$ into the Euler equation:

$$y = x^r$$

5. Substituting $y$ and dividing by $x^r$ leaves you with

$$2r^2 - 3r + 1 = 0$$

6. Factor the resulting characteristic equation as follows:

$$(r-1)(2r-1) = 0$$

7. So the roots are

$$r_1 = 1 \text{ and } r_2 = \frac{1}{2}$$

**11.** Determine the Euler equation that this differential equation is similar to; then determine the Euler equation's two roots, $r_1$ and $r_2$:

$$4x^2 \frac{d^2y}{dx^2} - 2x \frac{dy}{dx} + 2\left(1 + x^2\right)y = 0$$

*Solve It*

**12.** Find the Euler equation that this differential equation is similar to; then find the Euler equation's two roots, $r_1$ and $r_2$:

$$2x^2 \frac{d^2y}{dx^2} - 3x \frac{dy}{dx} + \left(2 + 2x + x^2\right)y = 0$$

*Solve It*

# Answers to Solving Differential Equations with Series Solutions Near Singular Points

Following are the answers to the practice questions presented throughout this chapter. Each one is worked out step by step so that if you messed one up along the way, you can more easily see where you took a wrong turn.

**1** **What are the singular points of this differential equation?**

$$x^2 \frac{d^2 y}{dx^2} - \left(8 - x^3\right) \frac{dy}{dx} + \left(1 - 9x\right) y = 0$$

**Solution:** $x_1 = 0$ and $x_2 = 0$

1. First, put the equation into the following form:

   $$\frac{d^2 y}{dx^2} + p(x) \frac{dy}{dx} + q(x) y = 0$$

   where

   $$p(x) = Q(x)/P(x)$$

   and

   $$q(x) = R(x)/P(x)$$

   Doing so gives you

   $$\frac{d^2 y}{dx^2} + \frac{\left(8 - x^3\right)}{x^2} \frac{dy}{dx} + \frac{\left(1 - 9x\right)}{x^2} y = 0$$

2. Therefore

   $$p(x) = \frac{\left(8 - x^3\right)}{x^2}$$

   and

   $$q(x) = \frac{\left(1 - 9x\right)}{x^2}$$

3. Looks like $p(x)$ and $q(x)$ both become unbounded when $x^2 = 0$, so the singular points are

   $$x_1 = 0 \text{ and } x_2 = 0$$

**2** **Solve for the singular points of this equation:**

$$\left(x - 9\right) \frac{d^2 y}{dx^2} - x^3 \frac{dy}{dx} + \frac{1}{x^2} y = 0$$

**Solution:** $x_1 = 9$ and $x_2 = 0$

1. Covert the differential equation to this form:

   $$\frac{d^2 y}{dx^2} + p(x) \frac{dy}{dx} + q(x) y = 0$$

   where

   $$p(x) = Q(x)/P(x)$$

   and

   $$q(x) = R(x)/P(x)$$

2. Now you have

$$\frac{d^2y}{dx^2} - \frac{x^3}{(x-9)}\frac{dy}{dx} + \frac{y}{(x-9)x^2} = 0$$

which means

$$p(x) = \frac{x^3}{(x-9)}$$

and

$$q(x) = \frac{1}{(x-9)x^2}$$

3. These answers tell you that $p(x)$ and $q(x)$ become unbounded when $x - 9 = 0$ (and that $q(x)$ also becomes unbounded when $x = 0$). Consequently, your singular points are

$x_1 = 9$ and $x_2 = 0$

**3** **What are the singular points of this differential equation?**

$$x^2\frac{d^2y}{dx^2} + x^3\frac{dy}{dx} + x^2y = 0$$

**Solution:** $x_1 = \infty$, $x_2 = -\infty$

1. First, put the equation into the following form:

$$\frac{d^2y}{dx^2} + p(x)\frac{dy}{dx} + q(x)y = 0$$

where

$p(x) = Q(x)/P(x)$

and

$q(x) = R(x)/P(x)$

Doing so gives you

$$\frac{d^2y}{dx^2} + x\frac{dy}{dx} + y = 0$$

2. Therefore

$p(x) = x$

and

$q(x) = 1$

3. Looks like $p(x)$ and $q(x)$ aren't unbounded anywhere, except when $x \to \pm\infty$, so the singular points are

$x_1 = \infty$ and $x_2 = -\infty$

**4** **Solve for the singular points of this equation:**

$$(x^2 + 3x + 2)\frac{d^2y}{dx^2} + \frac{dy}{dx} + y = 0$$

**Solution:** $x_1 = -1$, $x_2 = -2$

1. Convert the differential equation to this form:

$$\frac{d^2y}{dx^2} + p(x)\frac{dy}{dx} + q(x)y = 0$$

where

$$p(x) = Q(x)/P(x)$$

and

$$q(x) = R(x)/P(x)$$

2. Now you have

$$\frac{d^2y}{dx^2} + \frac{1}{\left(x^2 + 3x + 2\right)}\frac{dy}{dx} + \frac{y}{\left(x^2 + 3x + 2\right)} = 0$$

which means

$$p(x) = \frac{1}{\left(x^2 + 3x + 2\right)}$$

and

$$q(x) = \frac{1}{\left(x^2 + 3x + 2\right)}$$

3. These answers tell you that the roots of the denominators are −1 and −2. Consequently, your singular points are

$$x_1 = -1 \text{ and } x_2 = -2$$

**5**  **Are the singular points of this differential equation regular or irregular?**

$$x\frac{d^2y}{dx^2} + \frac{y}{x} = 0$$

**Solution: The singular points are regular.**

1. Start solving by putting the differential equation into the form

$$\frac{d^2y}{dx^2} + p(x)\frac{dy}{dx} + q(x)y = 0$$

where

$$p(x) = Q(x)/P(x)$$

and

$$q(x) = R(x)/P(x)$$

Doing so gives you

$$\frac{d^2y}{dx^2} + \frac{y}{x^2} = 0$$

2. Therefore

$$p(x) = 1$$

and

$$q(x) = \frac{1}{x^2}$$

So the singular points are $x_1 = 0$ and $x_2 = 0$.

3. The next step is to evaluate:

$$\lim_{x \to x_0}\left(x - x_0\right)p\left(x\right)$$

The singular points, $x_0$, are both 0, so this expression becomes

$$\lim_{x \to 0} x$$

and the value of the expression is 0, which is finite.

4. Nope, you're not done evaluating yet. Now you have to look at

$$\lim_{x \to x_0}\left(x - x_0\right)^2 q\left(x\right)$$

The singular points, $x_0$, are both 0, so this expression becomes

$$\lim_{x \to 0} 1$$

and the value of the expression is 1, which is finite.

5. Because the limits are finite, the singular points are regular.

**6** **Determine whether the singular points of this equation are regular or irregular:**

$$\frac{d^2y}{dx^2} + \frac{y}{\left(x^2 - 4\right)} = 0$$

**Solution: The singular points are regular.**

1. First things first. Convert the equation into the following form:

$$\frac{d^2y}{dx^2} + p\left(x\right)\frac{dy}{dx} + q\left(x\right)y = 0$$

where

$$p(x) = Q(x)/P(x)$$

and

$$q(x) = R(x)/P(x)$$

2. Now you have

$$\frac{d^2y}{dx^2} + \frac{y}{\left(x^2 - 4\right)} = 0$$

which means

$$p(x) = 1$$

and

$$q\left(x\right) = \frac{1}{\left(x^2 - 4\right)}$$

So the singular points are $x_1 = 2$ and $x_2 = -2$.

3. Evaluate:

$$\lim_{x \to x_0}\left(x - x_0\right)p\left(x\right)$$

The first singular point is 2, so this expression becomes

$$\lim_{x \to 2}\left(x - 2\right)$$

and the value of the expression is 0, which is finite.

The second singular point is $-2$, so that expression becomes

$$\lim_{x \to -2} \left( x + 2 \right)$$

and the value of the expression is 0, which is finite.

4. Next, evaluate:

$$\lim_{x \to x_0} \left( x - x_0 \right)^2 q\left( x \right)$$

The first singular point is 2, so the limit becomes

$$\lim_{x \to 2} \frac{\left( x - 2 \right)^2}{\left( x^2 - 4 \right)}$$

which equals

$$\lim_{x \to 2} \frac{\left( x - 2 \right)^2}{\left( x - 2 \right)\left( x + 2 \right)}$$

which in turn equals

$$\lim_{x \to 2} \frac{\left( x - 2 \right)}{\left( x + 2 \right)}$$

and the value of this expression is 0, which is finite.

The second singular point is $-2$, so the limit becomes

$$\lim_{x \to -2} \frac{\left( x + 2 \right)^2}{\left( x^2 - 4 \right)}$$

which equals

$$\lim_{x \to -2} \frac{\left( x + 2 \right)^2}{\left( x - 2 \right)\left( x + 2 \right)}$$

which in turn equals

$$\lim_{x \to -2} \frac{\left( x + 2 \right)}{\left( x - 2 \right)}$$

and the value of this expression is 0, which is finite.

5. The limits are finite, so guess what? The singular points are regular!

**7** **Find the solution to this differential equation:**

$$x^2 \frac{d^2 y}{dx^2} + 4x \, \frac{dy}{dx} + 2y = 0$$

**Solution:** $y = c_1 x^{-1} + c_2 x^{-2}$

1. This differential equation has the form of Euler's equation:

$$x^2 \frac{d^2 y}{dx^2} + \alpha x \, \frac{dy}{dx} + \beta y = 0$$

which means you can safely assume that its solution looks like

$$y = x^r$$

2. Substituting your attempted solution into the differential equation gives you

$$[r(r-1) + 4r + 2] \, x^r = 0$$

or

$$[r(r-1) + 4r + 2] = 0$$

which is actually

$$r^2 + 3r + 2 = 0$$

3. Factor that equation to get

$$(r+1)(r+2) = 0$$

4. It appears that the roots are

$$r_1 = -1 \text{ and } r_2 = -2$$

5. So the general solution to the original equation is

$$y = c_1 x^{-1} + c_2 x^{-2}$$

**8**    **Solve this differential equation:**

$$x^2 \frac{d^2 y}{dx^2} + 5x \, \frac{dy}{dx} + 3y = 0$$

**Solution:** $y = c_1 x^{-1} + c_2 x^{-3}$

1. Notice that the equation in the question has the form of Euler's equation:

$$x^2 \frac{d^2 y}{dx^2} + \alpha x \, \frac{dy}{dx} + \beta y = 0$$

2. Go ahead and assume that the solution is of this form:

$$y = x^r$$

3. Then substitute your attempted solution into the equation to get

$$[r(r-1) + 5r + 3] \, x^r = 0$$

or

$$[r(r-1) + 5r + 3] = 0$$

which is

$$r^2 + 4r + 3 = 0$$

4. Factoring this equation gives you

$$(r+1)(r+3) = 0$$

5. So the roots are

$$r_1 = -1 \text{ and } r_2 = -3$$

6. Therefore, the general solution to the differential equation is

$$y = c_1 x^{-1} + c_2 x^{-3}$$

**9**    **Find the solution to this differential equation:**

$$x^2 \frac{d^2 y}{dx^2} + 6x \, \frac{dy}{dx} + 6y = 0$$

**Solution:** $y = c_1 x^{-2} + c_2 x^{-3}$

1. This differential equation has the form of Euler's equation:

$$x^2 \frac{d^2 y}{dx^2} + \alpha x \frac{dy}{dx} + \beta y = 0$$

which means you can safely assume that its solution looks like

$$y = x^r$$

2. Substituting your attempted solution into the differential equation gives you

$$[r(r-1) + 6r + 6] \, x^r = 0$$

or

$$[r(r-1) + 6r + 6] = 0$$

which is actually

$$r^2 + 5r + 6 = 0$$

3. Factor that equation to get

$$(r+2)(r+3) = 0$$

4. It appears that the roots are

$$r_1 = -2 \text{ and } r_2 = -3$$

5. So the general solution to the differential equation is

$$y = c_1 x^{-2} + c_2 x^{-3}$$

**10** **Solve this differential equation:**

$$x^2 \frac{d^2 y}{dx^2} + 3x \frac{dy}{dx} + y = 0$$

**Solution:** $y = c_1 x^{-1} + c_2 \ln(x) \, x^{-1}$

1. Notice that the equation in the question has the form of Euler's equation:

$$x^2 \frac{d^2 y}{dx^2} + \alpha x \frac{dy}{dx} + \beta y = 0$$

2. Go ahead and assume that the solution is of this form:

$$y = x^r$$

3. Then substitute your attempted solution into the equation to get

$$[r(r-1) + 3r + 1] \, x^r = 0$$

or

$$[r(r-1) + 3r + 1] = 0$$

which is

$$r^2 + 2r + 1 = 0$$

4. Factoring this equation gives you

$$(r+1)(r+1) = 0$$

5. So the roots are

$$r_1 = -1 \text{ and } r_2 = -1$$

6. Therefore, the general solution to the differential equation is

$$y = c_1 x^{-1} + c_2 \ln (x) \, x^{-1}$$

**11** **Determine the Euler equation that this differential equation is similar to; then determine the Euler equation's two roots, $r_1$ and $r_2$:**

$$4x^2 \frac{d^2 y}{dx^2} - 2x \frac{dy}{dx} + 2\left(1 + x^2\right) y = 0$$

**Solution:** $x^2 \dfrac{d^2 y}{dx^2} - \dfrac{x}{2} \dfrac{dy}{dx} + \dfrac{y}{2} = 0$   **roots = 1, $\frac{1}{2}$**

1. First put the differential equation into this form:

$$x^2 \frac{d^2 y}{dx^2} - x \left[ xp(x) \right] \frac{dy}{dx} + \left[ x^2 q(x) \right] y = 0$$

Doing so gives you

$$p(x) = \frac{1}{2x}$$

and

$$q(x) = \frac{\left(1 + x^2\right)}{2x^2}$$

2. The Euler equation most similar to your differential equation is

$$x^2 \frac{d^2 y}{dx^2} - p_0 x \frac{dy}{dx} + q_0 y = 0$$

where

$$p_0 = \lim_{x \to 0} xp(x)$$

and

$$q_0 = \lim_{x \to 0} x^2 q(x)$$

so

$$p_0 = \frac{1}{2} \text{ and } q_0 = \frac{1}{2}$$

3. Therefore, the Euler equation is

$$x^2 \frac{d^2 y}{dx^2} - \frac{x}{2} \frac{dy}{dx} + \frac{y}{2} = 0$$

4. Multiplying by 2 gives you

$$2x^2 \frac{d^2 y}{dx^2} - x \frac{dy}{dx} + y = 0$$

5. To find the roots of the Euler equation, substitute this form of $y$ into the Euler equation:

$$y = x^r$$

6. Substituting $y$ and dividing by $x^r$ leaves you with

$$2r(r - 1) - r + 1 = 0$$

which you can expand into

$$2r^2 - 3r + 1 = 0$$

7. Factor the resulting characteristic equation as follows:

$$(r - 1)(2r - 1) = 0$$

8. So the roots are

$r_1 = 1$ and $r_2 = \frac{1}{2}$

**12** **Find the Euler equation that this differential equation is similar to; then find the Euler equation's two roots, $r_1$ and $r_2$:**

$$2x^2\frac{d^2y}{dx^2} - 3x\frac{dy}{dx} + \left(2 + 2x + x^2\right)y = 0$$

**Solution:** $2x^2\dfrac{d^2y}{dx^2} - 3x\dfrac{dy}{dx} + 2y = 0$   **roots = 2, $\frac{1}{2}$**

1. Begin by putting the differential equation into the following form:

$$x^2\frac{d^2y}{dx^2} - x\left[xp(x)\right]\frac{dy}{dx} + \left[x^2q(x)\right]y = 0$$

You now know that

$$p(x) = \frac{-3}{x}$$

and

$$q(x) = \frac{\left(2 + 2x + x^2\right)}{x^2}$$

2. The Euler equation most similar to your differential equation is

$$x^2\frac{d^2y}{dx^2} - p_0 x\frac{dy}{dx} + q_0 y = 0$$

where

$$p_0 = \lim_{x\to 0} xp(x)$$

and

$$q_0 = \lim_{x\to 0} x^2 q(x)$$

so

$p_0 = -3$ and $q_0 = 2$

3. Therefore, the Euler equation is

$$2x^2\frac{d^2y}{dx^2} - 3x\frac{dy}{dx} + 2y = 0$$

4. Substitute this form of $y$ into the Euler equation to find the equation's roots:

$y = x^r$

5. When you substitute $y$ and divide by $x^r$, you get

$2r(r-1) - 3r + 2 = 0$

which you can expand into

$2r^2 - 5r + 2 = 0$

6. Factoring the characteristic equation gives you

$(r-2)(2r-1) = 0$

7. Looks like the roots are

$r_1 = 2$ and $r_2 = \frac{1}{2}$

# Chapter 10

# Using Laplace Transforms to Solve Differential Equations

## In This Chapter

▶ Figuring out Laplace transforms by hand or by referencing a table

▶ Applying Laplace transforms when derivatives are in play

▶ Solving differential equations with the help of Laplace transforms

*L*aplace transforms, a type of integral transform, are another good tool in your differential equation solving toolkit. They have the great charm of being able to turn differential equations into algebra problems. Using algebra, you then group terms and see whether you have the recognizable Laplace transform of anything. If you do, you can get the reverse Laplace transform and your answer all in one fell swoop.

Care to see a Laplace transform in action? Take a standard differential equation like this one:

$$y'' + 5y' + 6y = 0$$

and find the Laplace transform of it, which looks like this (note that the $\mathcal{L}\{y\}$ term always indicates a Laplace transform):

$$\mathcal{L}\{y\} = \frac{1}{(s+2)} + \frac{1}{(s+3)}$$

From there you need to consult tables of Laplace transforms. If you can identify the Laplace transform of what you have, you're in business!

In this chapter, you practice finding Laplace transforms and then solving equations by using them.

## Finding Laplace Transforms

To unlock the magic of Laplace transforms, you need to be able to find the Laplace transform of the differential equation you're working with. That's why this section gives you practice calculating Laplace transforms of various mathematical expressions, such as exponentials and trigonometry functions.

Here's what a general integral transform looks like (note that this transform is not yet a Laplace transform):

$$F(s) = \int_{\alpha}^{\beta} K(s, t) f(t) \, dt$$

In this case, $f(t)$ is the function you're taking an integral transform of, and $F(s)$ is the transform. $K(s, t)$ is called the *kernel* of the transform, which is the function you mix into the integral. (When calculating a Laplace transform, you choose your own kernel because doing so gives you a chance to simplify your differential equation.) The limits of integration, $\alpha$ and $\beta$, can be anything you choose, but the most common limits for Laplace transforms are 0 to $+\infty$.

To calculate a Laplace transform by hand, simply follow these steps:

1. **Choose a kernel that transforms a differential equation into something simpler.**

2. **Try to invert the transform to get the solution of your original differential equation.**

When you restrict yourself to differential equations with constant coefficients, which is what you're doing in this chapter, a useful kernel is $e^{-st}$. Differentiating that with respect to $t$ gives you powers of $s$, which you can equate to the constant coefficients.

Here's what this handy kernel looks like when placed in the previous equation (note that besides using the kernel $e^{-st}$, the limits of integration are from 0 to $\infty$):

$$F(s) = \int_{0}^{\infty} e^{-st} f(t) \, dt$$

The symbol for Laplace transforms is $\mathcal{L}\{f(t)\}$, which is the Laplace transform of $f(t)$. So here's what the Laplace transform of the previous equation turns out to be:

$$\mathcal{L}\{f(t)\} = F(s) = \int_{0}^{\infty} e^{-st} f(t) \, dt$$

The good news is that you don't always have to go through the work of integrating by parts to find Laplace transforms. Sometimes you can use a table of Laplace transforms instead. To do so, simply take the Laplace transform of a differential equation and review a table of Laplace transforms (such as Table 10-1) to see whether you can identify any terms. Finding the inverse Laplace transform of the terms is easy: Just locate the correct entry in the table!

| Table 10-1 | Laplace Transforms of Common Functions | |
|---|---|---|
| *Function* | *Laplace Transform* | *Restrictions* |
| 1 | $\dfrac{1}{s}$ | $s > 0$ |
| $e^{st}$ | $\dfrac{1}{s-a}$ | $s > a$ |
| $t^n$ | $\dfrac{n!}{s^{n+1}}$ | $s > 0$, $n$ an integer $> 0$ |
| $\cos at$ | $\dfrac{s}{s^2 + a^2}$ | $s > 0$ |

| Function | Laplace Transform | Restrictions |
|---|---|---|
| sin $at$ | $\dfrac{a}{s^2+a^2}$ | $s>0$ |
| cosh $at$ | $\dfrac{s}{s^2-a^2}$ | $s>\|a\|$ |
| sinh $at$ | $\dfrac{a}{s^2-a^2}$ | $s>\|a\|$ |
| $e^{st}\cos bt$ | $\dfrac{s-a}{(s-a)^2+b^2}$ | $s>a$ |
| $e^{st}\sin bt$ | $\dfrac{b}{(s-a)^2+b^2}$ | $s>a$ |
| $t^n e^{st}$ | $\dfrac{n!}{(s-a)^{n+1}}$ | $s>a$, $n$ an integer $>0$ |
| $f(ct)$ | $\dfrac{1}{c}\mathcal{L}\{f(s/c)\}$ | $c>0$ |
| $f^{(n)}(t)$ | $s^n\mathcal{L}\{f(t)\}-s^{n-1}f(0)-...-sf^{(n-2)}(0)-f^{(n-1)}(0)$ | |

No table can possibly hold all the math expressions you may be asked to find the Laplace transform of, so be sure to practice finding Laplace transforms by hand from time to time.

Enough with all the review. Are you ready to find some Laplace transforms? Then check out the following problems and try to calculate them by hand (rather than relying on Table 10-1).

**Q.** What's the Laplace transform of sin ($at$)?

**A.** $\mathcal{L}\{\sin\ at\}=\dfrac{a}{s^2+a^2}\quad s>0$

1. Start with the general form of Laplace transforms and insert $e^{-st}$ as the kernel:

$$\mathcal{L}\{\sin\ at\}=\int_0^\infty \sin\ at\ e^{st}\,dt$$

2. Integrate by parts to get

$$\mathcal{L}\{\sin\ at\}=$$

$$\dfrac{e^{-st}\cos\ at}{a}\Big|_0^\infty -\dfrac{s}{a}\int_0^\infty \cos\ at\ e^{st}\,dt$$

which breaks down to

$$\mathcal{L}\{\sin\ at\}=\dfrac{1}{a}-\dfrac{s}{a}\int_0^\infty \cos\ at\ e^{st}\,dt$$

3. Note that the second term is similar to the original integral, except that it uses cosine. If you integrate by parts again,

you'll once again have an integral that uses sine:

$$\mathcal{L}\{\sin\ at\}=\dfrac{1}{a}-\dfrac{s^2}{a^2}\int_0^\infty \sin\ at\ e^{at}\,dt$$

4. The second term is actually $\dfrac{s^2}{a^2}$ multiplied by $\mathcal{L}\{\sin at\}$, which means the equation becomes

$$\mathcal{L}\{\sin\ at\}=\dfrac{1}{a}-\dfrac{s^2}{a^2}\mathcal{L}\{\sin\ at\}$$

5. You can recast this equation as follows:

$$\mathcal{L}\{\sin\ at\}+\dfrac{s^2}{a^2}\mathcal{L}\{\sin\ at\}=\dfrac{1}{a}$$

or

$$\dfrac{s^2+a^2}{a^2}\mathcal{L}\{\sin\ at\}=\dfrac{1}{a}$$

which becomes

$$\mathcal{L}\{\sin\ at\}=\dfrac{a}{s^2+a^2}\quad s>0$$

**1.** What's the Laplace transform of 1 (that is, $f(t) = 1$)?

$$\mathcal{L}\{1\} = ?$$

*Solve It*

**2.** Calculate the Laplace transform of $e^{at}$ (that is, $f(t) = e^{at}$):

$$\mathcal{L}\{e^{at}\} = ?$$

*Solve It*

# Calculating the Laplace Transforms of Derivatives

Occasionally, you're going to encounter differential equations such as the following that aren't so easy to take the Laplace transform of:

$$y'' + 5y' + 6y = 0$$

In order to find the Laplace transform of a derivative, all you have to do is follow this rule, which relates the Laplace transforms of derivatives:

$$\mathcal{L}\left\{y^{(n)}\right\} = s^n \mathcal{L}\left\{y\right\} - \sum_{k=1}^{n} s^{k-1} y^{(n-k)}(0)$$

Using this rule gives you

$$\mathcal{L}\{y''\} = s^2 \mathcal{L}\{y\} - sy(0) - y'(0)$$

and

$$\mathcal{L}\{y'\} = s \mathcal{L}\{y\} - y(0)$$

In the following example, I show you step by step how to find the Laplace transforms of derivatives. I then give you a couple chances to try the process out for yourself.

**Q.** What's the Laplace transform of $y'''$?

**A.** $\mathcal{L}\{y'''\} = s^3 \mathcal{L}\{y\} - y''(0) - sy'(0) - s^2 y(0)$

1. Use the rule that relates Laplace transforms of derivatives:
$$\mathcal{L}\left\{y^{(n)}\right\} = s^n \mathcal{L}\left\{y\right\} - \sum_{k=1}^{n} s^{k-1} y^{(n-k)}(0)$$

2. First, do the $k = 1$ term:
$$\mathcal{L}\left\{y'''\right\} = s^3 \mathcal{L}\left\{y\right\} - y''(0) \sum_{k=2}^{3} s^{k-1} y^{(n-k)}(0)$$

3. Then do the $k = 2$ term:
$$\mathcal{L}\left\{y'''\right\} = s^3 \mathcal{L}\left\{y\right\} - y''(0) -$$
$$sy'(0) - \sum_{k=3}^{3} s^{k-1} y^{(n-k)}(0)$$

4. Finally, do the $k = 3$ term:
$$\mathcal{L}\{y'''\} = s^3 \mathcal{L}\{y\} - y''(0) - sy'(0)$$
$$- s^2 y(0)$$

**3.** Find the Laplace transform of $y^{(4)}$:

*Solve It*

**4.** What's the Laplace transform of $y^{(5)}$?

*Solve It*

# Using Laplace Transforms to Solve Differential Equations

If you've reviewed the previous sections, you have all the tools you need to solve differential equations with the help of Laplace transforms. (If you haven't, I suggest you flip back a few pages and take a quick look at the earlier sections in this chapter.) The following practice problems let you put your skills to the test to solve various differential equations by using Laplace transforms. If you're feeling up to the challenge, skip straight to Question 5; otherwise, check out the following step-by-step example.

**Q.** Solve this differential equation by using Laplace transforms:

$$y'' + 4y' + 3y = 0$$

with the initial conditions

$$y(0) = 2$$

and

$$y'(0) = -4$$

**A.** $y = e^{-x} + e^{-3x}$

1. Take the Laplace transform of
   $$y'' + 4y' + 3y = 0$$
   to get
   $$\mathcal{L}\{y''\} + 4\mathcal{L}\{y'\} + 3\mathcal{L}\{y\}$$

2. Recall that the following equation is the Laplace transform of $y''$:
   $$\mathcal{L}\{y''\} = s^2\,\mathcal{L}\{y\} - sy(0) - y'(0)$$
   and that this equation is the Laplace transform of $y'$:
   $$\mathcal{L}\{y'\} = s\,\mathcal{L}\{y\} - y(0)$$

3. You wind up with this result for the differential equation:
   $$s^2\,\mathcal{L}\{y\} - sy(0) - y'(0) + 4[s\,\mathcal{L}\{y\} - y(0)] + 3\mathcal{L}\{y\} = 0$$

4. Collecting terms gives you
   $$(s^2 + 4s + 3)\,\mathcal{L}\{y\} - (4 + s)y(0) - y'(0) = 0$$

5. Now you can use the initial conditions
   $$y(0) = 2 \text{ and } y'(0) = -4$$
   to get
   $$(s^2 + 4s + 3)\,\mathcal{L}\{y\} - (8 + 2s) + 4 = 0$$
   or
   $$(s^2 + 4s + 3)\,\mathcal{L}\{y\} - 2s - 4 = 0$$

6. You now have
   $$\mathcal{L}\{y\} = \frac{2s + 4}{\left(s^2 + 4s + 3\right)}$$

7. Go ahead and factor the denominator so that you're left with this equation for the Laplace transform of the solution:
   $$\mathcal{L}\{y\} = \frac{2s + 4}{(s + 1)(s + 3)}$$

8. Your next step is to find a function whose Laplace transform is the preceding equation. To do that, use the method of partial fractions to get
   $$\mathcal{L}\{y\} = \frac{2s + 4}{(s + 1)(s + 3)} = \frac{a}{(s + 1)} + \frac{b}{(s + 3)}$$

9. Figure out what $a$ and $b$ are by writing your result as
   $$\mathcal{L}\{y\} = \frac{2s + 4}{(s + 1)(s + 3)} = \frac{a(s + 3) + b(s + 1)}{(s + 1)(s + 3)}$$

10. Then equate the numerators:
    $$2s + 4 = a(s + 3) + b(s + 1)$$

11. Because choosing $s$ is up to you, try setting it to $-1$ to get
    $$2 = 2a$$

which means that

$$1 = a$$

12. Now you can set $s$ to $-3$ to get

$$-2 = -2b$$

or

$$1 = b$$

13. So this equation:

$$\mathcal{L}\{y\} = \frac{a}{(s+1)} + \frac{b}{(s+3)}$$

becomes

$$\mathcal{L}\{y\} = \frac{1}{(s+1)} + \frac{1}{(s+3)}$$

Tada! You now know what the Laplace transform of the solution looks like. That means it's time to find the inverse Laplace transform of this equation.

14. By using Table 10-1, you can see that the Laplace transform of $e^{at}$ is

$$\mathcal{L}\{e^{a}\} = \frac{1}{s-a} \quad s > a$$

15. Compare $\mathcal{L}\{e^{at}\}$ to the first term in the Laplace transform of the solution, where $a = -1$, to find the solution's first term:

$$y_1 = e^{-x}$$

16. Then check out the second term in the Laplace transform of the solution:

$$\frac{1}{(s+3)}$$

17. Looks like the second term in the solution is

$$y_2 = e^{-3x}$$

18. Because the solution to the differential equation is $y = y_1 + y_2$, your result is

$$y = e^{-x} + e^{-3x}$$

**5.** Solve this differential equation by using Laplace transforms:

$$y'' + 3y' + 2y = 0$$

with the initial conditions

$$y(0) = 2$$

and

$$y'(0) = -3$$

Solve It

**6.** Using Laplace transforms, find the solution to this differential equation:

$$y'' + 5y' + 4y = 0$$

where

$$y(0) = 2$$

and

$$y'(0) = -5$$

Solve It

**7.** Solve this differential equation by using Laplace transforms:

$$y'' + 5y' + 6y = 0$$

with the initial conditions

$$y(0) = 2$$

and

$$y'(0) = -5$$

*Solve It*

**8.** Using Laplace transforms, find the solution to this differential equation:

$$y'' + 4y' + 3y = 0$$

where

$$y(0) = 3$$

and

$$y'(0) = -5$$

*Solve It*

**9.** Solve this differential equation by using Laplace transforms:

$y'' + 6y' + 5y = 0$

with the initial conditions

$y(0) = 5$

and

$y'(0) = -9$

*Solve It*

**10.** Using Laplace transforms, find the solution to this differential equation:

$y'' + 6y' + 8y = 0$

where

$y(0) = 4$

and

$y'(0) = -14$

*Solve It*

# Answers to Laplace Transform Problems

Here are the answers to the practice questions I provide throughout this chapter. I walk you through each answer so you can see the problems worked out step by step. Enjoy!

**1** **What's the Laplace transform of 1 (that is, $f(t) = 1$)?**

$$\mathcal{L}\{1\} = ?$$

**Solution:** $\mathcal{L}\{1\} = \dfrac{1}{s} \quad s > 0$

1. According to the definition of a Laplace transform:

$$\mathcal{L}\{1\} = \int_0^\infty e^{-st}(1)\ dt$$

or

$$\mathcal{L}\{1\} = \int_0^\infty e^{-st}dt$$

2. Performing the integration gives you

$$\mathcal{L}\{1\} = \int_0^\infty e^{-st}dt = -\frac{e^{-st}}{s}\bigg|_{t=0}^{t=\infty}$$

3. Okay. Now substitute in $t = 0$ and $t = \infty$:

$$\mathcal{L}\{1\} = 0 - \frac{-1}{s}$$

4. So $\mathcal{L}\{1\}$ equals

$$\mathcal{L}\{1\} = \frac{1}{s} \quad s > 0$$

and $\mathcal{L}\{1\}$ remains finite for all terms $s > 0$.

**2** **Calculate the Laplace transform of $e^{at}$ (that is, $f(t) = e^{at}$):**

$$\mathcal{L}\{e^{at}\} = ?$$

**Solution:** $\mathcal{L}\{e^{at}\} = \dfrac{1}{s-a} \quad s > a$

1. Here's what the Laplace transform of $e^{at}$ looks like:

$$\mathcal{L}\{e^{at}\} = \int_0^\infty e^{-st}e^{at}\ dt$$

2. This transform becomes

$$\mathcal{L}\{e^{at}\} = \int_0^\infty e^{-(s-a)t}\ dt$$

which in turn becomes

$$\mathcal{L}\{e^{at}\} = \int_0^\infty e^{-(s-a)t}\ dt = \frac{1}{s-a}e^{-(s-a)t}\bigg|_{t=0}^{t=\infty} \quad s > a$$

3. When you plug in the limits for $t$, you get

$$\mathcal{L}\{e^{at}\} = \int_0^\infty e^{-(s-a)t}dt = 0 - \frac{1}{s-a} \quad s > a$$

or

$$\mathcal{L}\{e^{at}\} = \frac{1}{s-a} \quad s > a$$

This result depends on the value you choose for $s$.

**3** Find the Laplace transform of $y^{(4)}$:

Solution: $\mathcal{L}\{y^{(4)}\} = s^4\,\mathcal{L}\{y\} - y'''(0) - sy''(0) - s^2 y'(0) - s^3 y(0)$

1. Use the rule that relates Laplace transforms of derivatives:

$$\mathcal{L}\left\{y^{(n)}\right\} = s^n \mathcal{L}\left\{y\right\} - \sum_{k=1}^{n} s^{k-1} y^{(n-k)}(0)$$

2. First, do the $k = 1$ term:

$$\mathcal{L}\left\{y^{(4)}\right\} = s^4 \mathcal{L}\left\{y\right\} - y'''(0) - \sum_{k=2}^{4} s^{k-1} y^{(n-k)}(0)$$

3. Then do the $k = 2$ term:

$$\mathcal{L}\left\{y^{(4)}\right\} = s^4 \mathcal{L}\left\{y\right\} - y'''(0) - sy''(0) - \sum_{k=3}^{4} s^{k-1} y^{(n-k)}(0)$$

4. Next, do the $k = 3$ term:

$$\mathcal{L}\left\{y^{(4)}\right\} = s^4 \mathcal{L}\left\{y\right\} - y'''(0) - sy''(0) - s^2 y'(0) - \sum_{k=4}^{4} s^{k-1} y^{(n-k)}(0)$$

5. Finally, do the $k = 4$ term and put everything together:

$$\mathcal{L}\{y^{(4)}\} = s^4\,\mathcal{L}\{y\} - y'''(0) - sy''(0) - s^2 y'(0) - s^3 y(0)$$

**4** What's the Laplace transform of $y^{(5)}$?

Solution: $\mathcal{L}\{y^{(5)}\} = s^5\,\mathcal{L}\{y\} - y^{(4)}(0) - sy'''(0) - s^2 y''(0) - s^3 y'(0) - s^4 y(0)$

1. Use the rule that relates Laplace transforms of derivatives:

$$\mathcal{L}\left\{y^{(n)}\right\} = s^n \mathcal{L}\left\{y\right\} - \sum_{k=1}^{n} s^{k-1} y^{(n-k)}(0)$$

2. Start applying the rule by doing the $k = 1$ term:

$$\mathcal{L}\left\{y^{(5)}\right\} = s^5 \mathcal{L}\left\{y\right\} - y^{(4)}(0) - \sum_{k=2}^{5} s^{k-1} y^{(n-k)}(0)$$

3. Then do the $k = 2$ term:

$$\mathcal{L}\left\{y^{(5)}\right\} = s^5 \mathcal{L}\left\{y\right\} - y^{(4)}(0) - sy'''(0) - \sum_{k=3}^{5} s^{k-1} y^{(n-k)}(0)$$

4. You guessed it, do the $k = 3$ term next:

$$\mathcal{L}\left\{y^{(5)}\right\} = s^5 \mathcal{L}\left\{y\right\} - y^{(4)}(0) - sy'''(0) - s^2 y''(0) - \sum_{k=4}^{5} s^{k-1} y^{(n-k)}(0)$$

5. Then do the $k = 4$ term:

$$\mathcal{L}\left\{y^{(5)}\right\} = s^5 \mathcal{L}\left\{y\right\} - y^{(4)}(0) - sy'''(0) - s^2 y''(0) - s^3 y'(0) - \sum_{k=5}^{5} s^{k-1} y^{(n-k)}(0)$$

6. Finally, do the $k = 5$ term and put everything together:

$$\mathcal{L}\{y^{(5)}\} = s^5\,\mathcal{L}\{y\} - y^{(4)}(0) - sy'''(0) - s^2 y''(0) - s^3 y'(0) - s^4 y(0)$$

**5** Solve this differential equation by using Laplace transforms:

$$y'' + 3y' + 2y = 0$$

with the initial conditions

$$y(0) = 2$$

and

$$y'(0) = -3$$

Solution: $y = e^{-x} + e^{-2x}$

1. Take the Laplace transform of

   $y'' + 3y' + 2y = 0$

   to get

   $\mathcal{L}\{y''\} + 3\mathcal{L}\{y'\} + 2\mathcal{L}\{y\}$

2. Recall that the following equation is the Laplace transform of $y''$:

   $\mathcal{L}\{y''\} = s^2\mathcal{L}\{y\} - sy(0) - y'(0)$

   and that this equation is the Laplace transform of $y'$:

   $\mathcal{L}\{y'\} = s\,\mathcal{L}\{y\} - y(0)$

3. You wind up with this result for the differential equation:

   $s^2\mathcal{L}\{y\} - sy(0) - y'(0) + 3[s\,\mathcal{L}\{y\} - y(0)] + 2\mathcal{L}\{y\} = 0$

4. Collecting terms gives you

   $(s^2 + 3s + 2)\mathcal{L}\{y\} - (3 + s)y(0) - y'(0) = 0$

5. Now you can use the initial conditions

   $y(0) = 2$ and $y'(0) = -3$

   to get

   $(s^2 + 3s + 2)\mathcal{L}\{y\} - (6 + 2s) + 3 = 0$

   or

   $(s^2 + 3s + 2)\mathcal{L}\{y\} - 2s - 3 = 0$

6. You now have

   $$\mathcal{L}\{y\} = \frac{2s+3}{\left(s^2+3s+2\right)}$$

7. Go ahead and factor the denominator so that you're left with this equation for the Laplace transform of the solution:

   $$\mathcal{L}\{y\} = \frac{2s+3}{\left(s+1\right)\left(s+2\right)}$$

8. Your next step is to find a function whose Laplace transform is the preceding equation. To do that, use the method of partial fractions to get

   $$\mathcal{L}\{y\} = \frac{2s+3}{\left(s+1\right)\left(s+2\right)} = \frac{a}{\left(s+1\right)} + \frac{b}{\left(s+2\right)}$$

9. Figure out what $a$ and $b$ are by writing your result as

   $$\mathcal{L}\{y\} = \frac{2s+3}{\left(s+1\right)\left(s+3\right)} = \frac{a\left(s+2\right)+b\left(s+1\right)}{\left(s+1\right)\left(s+2\right)}$$

10. Then equate the numerators:

    $2s + 3 = a(s + 2) + b(s + 1)$

11. Because choosing $s$ is up to you, try setting it to $-1$ to get

    $2 = 2a$

    which means that

    $1 = a$

12. Now you can set $s$ to $-2$ to get

$$-1 = -b$$

or

$$1 = b$$

13. So this equation:

$$\mathcal{L}\{y\} = \frac{a}{(s+1)} + \frac{b}{(s+2)}$$

becomes

$$\mathcal{L}\{y\} = \frac{1}{(s+1)} + \frac{1}{(s+2)}$$

Tada! You now know what the Laplace transform of the solution looks like. That means it's time to find the inverse Laplace transform of this equation.

14. By using Table 10-1, you can see that the Laplace transform of $e^{at}$ is

$$\mathcal{L}\{e^a\} = \frac{1}{s-a} \quad s > a$$

15. Compare $\mathcal{L}\{e^{at}\}$ to the first term in the Laplace transform of the solution, where $a = -1$, to find the solution's first term:

$$y_1 = e^{-x}$$

16. Then check out the second term in the Laplace transform of the solution:

$$\frac{1}{(s+2)}$$

17. Looks like the second term in the solution is

$$y_2 = e^{-2x}$$

18. Because the solution to the differential equation is $y = y_1 + y_2$, your result is

$$y = e^{-x} + e^{-2x}$$

**6** **Using Laplace transforms, find the solution to this differential equation:**

$$y'' + 5y' + 4y = 0$$

**where**

$$y(0) = 2$$

**and**

$$y'(0) = -5$$

**Solution:** $y = e^{-x} + e^{-4x}$

1. Find the Laplace transform of the differential equation:

$$\mathcal{L}\{y''\} + 5\mathcal{L}\{y'\} + 4\mathcal{L}\{y\}$$

2. Recall that this equation is the Laplace transform of $y''$:

$$\mathcal{L}\{y''\} = s^2 \mathcal{L}\{y\} - sy(0) - y'(0)$$

and that this equation is the Laplace transform of $y'$:

$$\mathcal{L}\{y'\} = s \mathcal{L}\{y\} - y(0)$$

3. Consequently, your result for the differential equation is

$$s^2 \mathcal{L}\{y\} - sy(0) - y'(0) + 5[s\,\mathcal{L}\{y\} - y(0)] + 4\mathcal{L}\{y\} = 0$$

4. Collect terms to get

$$(s^2 + 5s + 4)\,\mathcal{L}\{y\} - (5 + s)y(0) - y'(0) = 0$$

5. At long last, you can use the initial conditions

$$y(0) = 2 \text{ and } y'(0) = -5$$

which give you

$$(s^2 + 5s + 4)\,\mathcal{L}\{y\} - (10 + 2s) + 5 = 0$$

or

$$(s^2 + 5s + 4)\,\mathcal{L}\{y\} - 2s - 5 = 0$$

6. Now you have the following:

$$\mathcal{L}\{y\} = \frac{2s+5}{\left(s^2+5s+4\right)}$$

7. Factoring the denominator leaves you with this equation for the Laplace transform of the solution:

$$\mathcal{L}\{y\} = \frac{2s+5}{\left(s+1\right)\left(s+4\right)}$$

8. Next, use the method of partial fractions to find a function whose Laplace transform is the equation shown in Step 7:

$$\mathcal{L}\{y\} = \frac{2s+5}{\left(s+1\right)\left(s+4\right)} = \frac{a}{\left(s+1\right)} + \frac{b}{\left(s+4\right)}$$

9. To figure out what $a$ and $b$ are, write your result as

$$\mathcal{L}\{y\} = \frac{2s+5}{\left(s+1\right)\left(s+4\right)} = \frac{a\left(s+4\right)+b\left(s+1\right)}{\left(s+1\right)\left(s+4\right)}$$

10. Now equate the numerators to get

$$2s + 5 = a(s + 4) + b(s + 1)$$

11. Then try setting $s$ to $-1$. You wind up with

$$3 = 3a$$

which means that

$$1 = a$$

12. Now you can set $s$ to $-4$ to get

$$-3 = -3b$$

or

$$1 = b$$

13. So this equation:

$$\mathcal{L}\{y\} = \frac{a}{\left(s+1\right)} + \frac{b}{\left(s+4\right)}$$

becomes

$$\mathcal{L}\{y\} = \frac{1}{(s+1)} + \frac{1}{(s+4)}$$

That's what the Laplace transform of the solution looks like.

14. To find the inverse Laplace transform of the preceding equation, refer to Table 10-1. There you can see that the Laplace transform of $e^{at}$ is

$$\mathcal{L}\{e^a\} = \frac{1}{s-a} \quad s > a$$

15. Comparing $\mathcal{L}\{e^{at}\}$ to the first term in the Laplace transform of the solution, where $a = -1$, shows you that the first term in the solution is

$$y_1 = e^{-x}$$

16. A quick look at the second term in the Laplace transform of the solution

$$\frac{1}{(s+4)}$$

reveals that the solution's second term is

$$y_2 = e^{-4x}$$

17. Thus, the solution to the differential equation is $y = y_1 + y_2$, which equals

$$y = e^{-x} + e^{-4x}$$

**7** **Solve this differential equation by using Laplace transforms:**

$$y'' + 5y' + 6y = 0$$

**with the initial conditions**

$$y(0) = 2$$

**and**

$$y'(0) = -5$$

**Solution:** $y = e^{-2x} + e^{-3x}$

1. Take the Laplace transform of

$$y'' + 5y' + 6y = 0$$

to get

$$\mathcal{L}\{y''\} + 5\mathcal{L}\{y'\} + 6\mathcal{L}\{y\}$$

2. Recall that the following equation is the Laplace transform of $y''$:

$$\mathcal{L}\{y''\} = s^2\,\mathcal{L}\{y\} - sy(0) - y'(0)$$

and that this equation is the Laplace transform of $y'$:

$$\mathcal{L}\{y'\} = s\,\mathcal{L}\{y\} - y(0)$$

3. You wind up with this result for the differential equation:

$$s^2\,\mathcal{L}\{y\} - sy(0) - y'(0) + 5[s\,\mathcal{L}\{y\} - y(0)] + 6\mathcal{L}\{y\} = 0$$

4. Collecting terms gives you

$$(s^2 + 5s + 6)\,\mathcal{L}\{y\} - (5 + s)y(0) - y'(0) = 0$$

5. Now you can use the initial conditions

    $y(0) = 2$ and $y'(0) = -5$

    to get

    $(s^2 + 5s + 6)\,\mathcal{L}\{y\} - (10 + 2s) + 5 = 0$

    or

    $(s^2 + 5s + 6)\,\mathcal{L}\{y\} - 2s - 5 = 0$

6. You now have

    $$\mathcal{L}\{y\} = \frac{2s+5}{\left(s^2+5s+6\right)}$$

7. Go ahead and factor the denominator so that you're left with this equation for the Laplace transform of the solution:

    $$\mathcal{L}\{y\} = \frac{2s+5}{(s+2)(s+3)}$$

8. Your next step is to find a function whose Laplace transform is the preceding equation. To do that, use the method of partial fractions to get

    $$\mathcal{L}\{y\} = \frac{2s+5}{(s+2)(s+3)} = \frac{a}{(s+2)} + \frac{b}{(s+3)}$$

9. Figure out what $a$ and $b$ are by writing your result as

    $$\mathcal{L}\{y\} = \frac{2s+5}{(s+2)(s+3)} = \frac{a(s+2)+b(s+3)}{(s+2)(s+3)}$$

10. Then equate the numerators:

    $2s + 5 = a(s + 2) + b(s + 3)$

11. Because choosing $s$ is up to you, try setting it to $-3$ to get

    $-1 = -a$

    which means that

    $1 = a$

12. Now you can set $s$ to $-2$ to get

    $1 = b$

13. So this equation:

    $$\mathcal{L}\{y\} = \frac{a}{(s+2)} + \frac{b}{(s+3)}$$

    becomes

    $$\mathcal{L}\{y\} = \frac{1}{(s+2)} + \frac{1}{(s+3)}$$

    Tada! You now know what the Laplace transform of the solution looks like. That means it's time to find the inverse Laplace transform of this equation.

14. By using Table 10-1, you can see that the Laplace transform of $e^{at}$ is

    $$\mathcal{L}\{e^{a}\} = \frac{1}{s-a} \quad s > a$$

15. Compare $\mathcal{L}\{e^{at}\}$ to the first term in the Laplace transform of the solution, where $a = -2$, to find the solution's first term:

$$y_1 = e^{-2x}$$

16. Then check out the second term in the Laplace transform of the solution:

$$\frac{1}{(s+3)}$$

17. Looks like the second term in the solution is

$$y_2 = e^{-3x}$$

18. Because the solution to the differential equation is $y = y_1 + y_2$, your result is

$$y = e^{-2x} + e^{-3x}$$

**8** **Using Laplace transforms, find the solution to this differential equation:**

$$y'' + 4y' + 3y = 0$$

**where**

$$y(0) = 3$$

**and**

$$y'(0) = -5$$

**Solution:** $y = 2e^{-x} + e^{-3x}$

1. Find the Laplace transform of the differential equation:

$$\mathcal{L}\{y''\} + 4\mathcal{L}\{y'\} + 3\mathcal{L}\{y\}$$

2. Recall that this equation is the Laplace transform of $y''$:

$$\mathcal{L}\{y''\} = s^2 \mathcal{L}\{y\} - sy(0) - y'(0)$$

and that this equation is the Laplace transform of $y'$:

$$\mathcal{L}\{y'\} = s \mathcal{L}\{y\} - y(0)$$

3. Consequently, your result for the differential equation is

$$s^2 \mathcal{L}\{y\} - sy(0) - y'(0) + 4[s \mathcal{L}\{y\} - y(0)] + 3\mathcal{L}\{y\} = 0$$

4. Collect terms to get

$$(s^2 + 4s + 3) \mathcal{L}\{y\} - (4 + s)y(0) - y'(0) = 0$$

5. At long last, you can use the initial conditions

$$y(0) = 3 \text{ and } y'(0) = -5$$

which give you

$$(s^2 + 4s + 3) \mathcal{L}\{y\} - (12 + 3s) + 5 = 0$$

or

$$(s^2 + 4s + 3) \mathcal{L}\{y\} - 3s - 7 = 0$$

6. Now you have the following:

$$\mathcal{L}\{y\} = \frac{3s+7}{(s^2 + 4s + 3)}$$

7. Factoring the denominator leaves you with this equation for the Laplace transform of the solution:

$$\mathcal{L}\{y\} = \frac{3s+7}{(s+1)(s+3)}$$

8. Next, use the method of partial fractions to find a function whose Laplace transform is the equation shown in Step 7:

$$\mathcal{L}\{y\} = \frac{3s+7}{(s+1)(s+3)} = \frac{a}{(s+1)} + \frac{b}{(s+3)}$$

9. To figure out what $a$ and $b$ are, write your result as

$$\mathcal{L}\{y\} = \frac{3s+7}{(s+1)(s+3)} = \frac{a(s+3)+b(s+1)}{(s+1)(s+3)}$$

10. Now equate the numerators to get

$$3s + 7 = a(s + 3) + b(s + 1)$$

11. Then try setting $s$ to $-1$. You wind up with

$$4 = 2a$$

which means that

$$2 = a$$

12. Now you can set $s$ to $-3$ to get

$$-2 = -2b$$

or

$$1 = b$$

13. So this equation:

$$\mathcal{L}\{y\} = \frac{a}{(s+1)} + \frac{b}{(s+3)}$$

becomes

$$\mathcal{L}\{y\} = \frac{2}{(s+1)} + \frac{1}{(s+3)}$$

That's what the Laplace transform of the solution looks like.

14. To find the inverse Laplace transform of the preceding equation, refer to Table 10-1. There you can see that the Laplace transform of $e^{at}$ is

$$\mathcal{L}\{e^{a}\} = \frac{1}{s-a} \quad s > a$$

15. Comparing $\mathcal{L}\{e^{at}\}$ to the first term in the Laplace transform of the solution, where $a = -1$, shows you that the first term in the solution is

$$y_1 = 2e^{-x}$$

16. A quick look at the second term in the Laplace transform of the solution

$$\frac{1}{(s+3)}$$

reveals that the solution's second term is

$$y_2 = e^{-3x}$$

17. Thus, the solution to the differential equation is $y = y_1 + y_2$, which equals

$$y = 2e^{-x} + e^{-3x}$$

**9** Solve this differential equation by using Laplace transforms:

$$y'' + 6y' + 5y = 0$$

**with the initial conditions**

$$y(0) = 5$$

**and**

$$y'(0) = -9$$

Solution: $y = 4e^{-x} + e^{-5x}$

1. Take the Laplace transform of

$$y'' + 6y' + 5y = 0$$

to get

$$\mathcal{L}\{y''\} + 6\mathcal{L}\{y'\} + 5\mathcal{L}\{y\}$$

2. Recall that the following equation is the Laplace transform of $y''$:

$$\mathcal{L}\{y''\} = s^2 \mathcal{L}\{y\} - sy(0) - y'(0)$$

and that this equation is the Laplace transform of $y'$:

$$\mathcal{L}\{y'\} = s \mathcal{L}\{y\} - y(0)$$

3. You wind up with this result for the differential equation:

$$s^2 \mathcal{L}\{y\} - sy(0) - y'(0) + 6[s \mathcal{L}\{y\} - y(0)] + 5\mathcal{L}\{y\} = 0$$

4. Collecting terms gives you

$$(s^2 + 6s + 5) \mathcal{L}\{y\} - (6 + s)y(0) - y'(0) = 0$$

5. Now you can use the initial conditions

$$y(0) = 5 \text{ and } y'(0) = -9$$

to get

$$(s^2 + 6s + 5) \mathcal{L}\{y\} - (30 + 5s) + 9 = 0$$

or

$$(s^2 + 6s + 5) \mathcal{L}\{y\} - 5s - 21 = 0$$

6. You now have

$$\mathcal{L}\{y\} = \frac{5s + 21}{\left(s^2 + 6s + 5\right)}$$

7. Go ahead and factor the denominator so that you're left with this equation for the Laplace transform of the solution:

$$\mathcal{L}\{y\} = \frac{5s + 21}{\left(s + 1\right)\left(s + 5\right)}$$

8. Your next step is to find a function whose Laplace transform is the preceding equation. To do that, use the method of partial fractions to get

$$\mathcal{L}\{y\} = \frac{5s+21}{(s+1)(s+5)} = \frac{a}{(s+1)} + \frac{b}{(s+5)}$$

9. Figure out what $a$ and $b$ are by writing your result as

$$\mathcal{L}\{y\} = \frac{5s+21}{(s+1)(s+5)} = \frac{a(s+5)+b(s+1)}{(s+1)(s+5)}$$

10. Then equate the numerators:

$$5s + 21 = a(s + 5) + b(s + 1)$$

11. Because choosing $s$ is up to you, try setting it to $-1$ to get

$$16 = 4a$$

which means that

$$4 = a$$

12. Now you can set $s$ to $-5$ to get

$$-4 = -4b$$

or

$$1 = b$$

13. So this equation:

$$\mathcal{L}\{y\} = \frac{a}{(s+1)} + \frac{b}{(s+3)}$$

becomes

$$\mathcal{L}\{y\} = \frac{4}{(s+1)} + \frac{1}{(s+3)}$$

Tada! You now know what the Laplace transform of the solution looks like. That means it's time to find the inverse Laplace transform of this equation.

14. By using Table 10-1, you can see that the Laplace transform of $e^{at}$ is

$$\mathcal{L}\{e^{a}\} = \frac{1}{s-a} \quad s > a$$

15. Compare $\mathcal{L}\{e^{at}\}$ to the first term in the Laplace transform of the solution, where $a = -1$, to find the solution's first term:

$$y_1 = 4e^{-x}$$

16. Then check out the second term in the Laplace transform of the solution:

$$\frac{1}{(s+3)}$$

17. Looks like the second term in the solution is

$$y_2 = e^{-5x}$$

18. Because the solution to the differential equation is $y = y_1 + y_2$, your final result is

$$y = 4e^{-x} + e^{-5x}$$

**10** **Using Laplace transforms, find the solution to this differential equation:**

$$y'' + 6y' + 8y = 0$$

**where**

$$y(0) = 4$$

**and**

$$y'(0) = -14$$

**Solution:** $y = e^{-2x} + 3e^{-4x}$

1. Find the Laplace transform of the differential equation:

$$\mathcal{L}\{y''\} + 6\mathcal{L}\{y'\} + 8\mathcal{L}\{y\}$$

2. Recall that this equation is the Laplace transform of $y''$:

$$\mathcal{L}\{y''\} = s^2\mathcal{L}\{y\} - sy(0) - y'(0)$$

and that this equation is the Laplace transform of $y'$:

$$\mathcal{L}\{y'\} = s\mathcal{L}\{y\} - y(0)$$

3. Consequently, your result for the differential equation is

$$s^2\mathcal{L}\{y\} - sy(0) - y'(0) + 6[s\mathcal{L}\{y\} - y(0)] + 8\mathcal{L}\{y\} = 0$$

4. Collect terms to get

$$(s^2 + 6s + 8)\mathcal{L}\{y\} - (6 + s)y(0) - y'(0) = 0$$

5. At long last, you can use the initial conditions

$$y(0) = 4 \text{ and } y'(0) = -14$$

which give you

$$(s^2 + 6s + 8)\mathcal{L}\{y\} - (24 + 4s) + 14 = 0$$

or

$$(s^2 + 6s + 8)\mathcal{L}\{y\} - 4s - 10 = 0$$

6. Now you have the following:

$$\mathcal{L}\{y\} = \frac{4s + 10}{(s^2 + 6s + 8)}$$

7. Factoring the denominator leaves you with this equation for the Laplace transform of the solution:

$$\mathcal{L}\{y\} = \frac{4s + 10}{(s + 2)(s + 4)}$$

8. Next, use the method of partial fractions to find a function whose Laplace transform is the equation shown in Step 7:

$$\mathcal{L}\{y\} = \frac{4s + 10}{(s + 2)(s + 4)} = \frac{a}{(s + 2)} + \frac{b}{(s + 4)}$$

9. To figure out what $a$ and $b$ are, write your result as

$$\mathcal{L}\{y\} = \frac{4s+10}{(s+2)(s+4)} = \frac{a(s+4)+b(s+2)}{(s+2)(s+4)}$$

10. Now equate the numerators to get

$$4s + 10 = a(s + 4) + b(s + 2)$$

11. Then try setting $s$ to $-2$. You wind up with

$$2 = 2a$$

which means that

$$1 = a$$

12. Now you can set $s$ to $-4$ to get

$$-6 = -2b$$

or

$$3 = b$$

13. So this equation:

$$\mathcal{L}\{y\} = \frac{a}{(s+2)} + \frac{b}{(s+4)}$$

becomes

$$\mathcal{L}\{y\} = \frac{1}{(s+2)} + \frac{3}{(s+4)}$$

That's what the Laplace transform of the solution looks like.

14. To find the inverse Laplace transform of the preceding equation, refer to Table 10-1. There you can see that the Laplace transform of $e^{at}$ is

$$\mathcal{L}\{e^{a}\} = \frac{1}{s-a} \quad s > a$$

15. Comparing $\mathcal{L}\{e^{at}\}$ to the first term in the Laplace transform of the solution, where $a = -1$, shows you that the first term in the solution is

$$y_1 = e^{-2x}$$

16. A quick look at the second term in the Laplace transform of the solution

$$\frac{3}{(s+4)}$$

reveals that the solution's second term is

$$y_2 = 3e^{-4x}$$

17. Thus, the solution to the differential equation is $y = y_1 + y_2$, which equals

$$y = e^{-2x} + 3e^{-4x}$$

# Chapter 11

# Solving Systems of Linear First Order Differential Equations

• • • • • • • • • • • • • • • • • • • • • • • • • • • • • • • • • • • • • • • • • • • • • • • • • • • • •

*In This Chapter*

▶ Reviewing the basics of matrix operations

▶ Calculating the determinant

▶ Figuring out eigenvalues and eigenvectors

▶ Seeking out the solution to various systems

• • • • • • • • • • • • • • • • • • • • • • • • • • • • • • • • • • • • • • • • • • • • • • • • • • • • •

*T*his chapter is all about solving systems of linear first order differential equations and practicing various techniques involved in solving systems. First, you work out your matrix-handling skills by adding matrices, multiplying them, and finding their determinants. Then you practice finding eigenvalues and eigenvectors. Finally — drum roll please — you tackle solving some systems of differential equations.

# *Back to the Basics: Adding (And Subtracting) Matrices*

Before you can solve systems of linear first order differential equations with the aid of matrices, you need to know how to handle some basic matrix operations, starting with the most basic of all: addition and subtraction.

Adding two matrices together involves adding the elements at corresponding positions in the two matrices. The same is true when subtracting matrices: Elements must be at corresponding positions in order for you to subtract them.

When adding matrices, such as **A** and **B**, you can flip the order of the matrices and still get the same solution. In other words, **A** + **B** = **B** + **A**. However, subtracting matrices doesn't work the same way. **A** – **B** can't possibly equal **B** – **A**. In fact, when you flip the matrices within the operation, **A** – **B** = –(**B** – **A**).

Following is an example problem to refresh your memory on adding matrices. After reviewing it, check out the following questions for some practice adding matrices. (Where are the subtraction practice problems, you ask? The process of adding and subtracting matrices is so similar that I've spared you the hassle of working on subtraction here so you can focus your energy on solving the differential equations presented later in this chapter.)

**Q.** What's **A** + **B** if **A** and **B** are

$$A = \begin{pmatrix} 1 & 1 \\ 1 & 1 \end{pmatrix} \text{ and } B = \begin{pmatrix} 2 & 2 \\ 2 & 2 \end{pmatrix}$$

**A.** $A + B = \begin{pmatrix} 3 & 3 \\ 3 & 3 \end{pmatrix}$

1. **A** + **B** is

$$\begin{pmatrix} 1 & 1 \\ 1 & 1 \end{pmatrix} + \begin{pmatrix} 2 & 2 \\ 2 & 2 \end{pmatrix}$$

2. By adding element to element, you get

$$\begin{pmatrix} 1 & 1 \\ 1 & 1 \end{pmatrix} + \begin{pmatrix} 2 & 2 \\ 2 & 2 \end{pmatrix} = \begin{pmatrix} 3 & 3 \\ 3 & 3 \end{pmatrix}$$

**1.** Find the sum of **A** + **B** if **A** and **B** are

$$A = \begin{pmatrix} 4 & 3 \\ 2 & 1 \end{pmatrix} \text{ and } B = \begin{pmatrix} 1 & 2 \\ 3 & 4 \end{pmatrix}$$

*Solve It*

**2.** What's **A** + **B** if **A** and **B** are

$$A = \begin{pmatrix} 9 & 9 \\ 9 & 9 \end{pmatrix} \text{ and } B = \begin{pmatrix} 1 & 2 \\ 3 & 4 \end{pmatrix}$$

*Solve It*

# An Exercise in Muddying Your Mind: Multiplying Matrices

To effectively employ matrices when solving systems of equations, you need to have some matrix-multiplication skills in your arsenal. Multiplying matrices is a little more involved than simply adding them. Why? Because **AB** is defined when the number of columns in **A** is the same as the number of rows in **B**. That is, if **A** is an $l \times m$ (that's row $\times$ column notation, so **A** has $l$ rows and $m$ columns) matrix and **B** is an $m \times n$ matrix, then the product **AB** exists — and the product is an $l \times n$ matrix.

If **AB** = **C**, then the $(i, j)$ (that's row, column) element of **C** is found by multiplying each element of the $i$th row of **A** by the matching element in the $j$th column of **B** and then adding the resulting products. Following is the standard visual presentation of multiplying matrices:

$$\mathbf{C}_{ij} = \sum_{k=1}^{m} \mathbf{A}_{ik} \mathbf{B}_{kj}$$

Here's a little tidbit that may come in handy for you: **AB** $\neq$ **BA**.

As you multiply matrices, you may occasionally encounter something called the *identity matrix*. It's labeled **I** and holds 1s along its upper-left to lower-right diagonal; the other numbers in it are all 0s. Check out this $2 \times 2$ identity matrix to see what I mean:

$$\mathbf{I} = \begin{pmatrix} 1 & 0 \\ 0 & 1 \end{pmatrix}$$

A $3 \times 3$ identity matrix looks like this:

$$\mathbf{I} = \begin{pmatrix} 1 & 0 & 0 \\ 0 & 1 & 0 \\ 0 & 0 & 1 \end{pmatrix}$$

Multiplying any matrix, **A,** by the identity matrix gives you **A** back again. For example, take a look at this multiplication:

$$\mathbf{AI} = \begin{pmatrix} 1 & 2 & 3 \\ 4 & 5 & 6 \\ 7 & 8 & 9 \end{pmatrix}\begin{pmatrix} 1 & 0 & 0 \\ 0 & 1 & 0 \\ 0 & 0 & 1 \end{pmatrix}$$

The product of **AI** is just **A** all by itself:

$$\mathbf{AI} = \begin{pmatrix} 1 & 2 & 3 \\ 4 & 5 & 6 \\ 7 & 8 & 9 \end{pmatrix}\begin{pmatrix} 1 & 0 & 0 \\ 0 & 1 & 0 \\ 0 & 0 & 1 \end{pmatrix} = \begin{pmatrix} 1 & 2 & 3 \\ 4 & 5 & 6 \\ 7 & 8 & 9 \end{pmatrix}$$

So are you ready to practice multiplying matrices? Well, here's your chance!

**Q.** What's the product of **A** and **B** if

$$A = \begin{pmatrix} 1 & 2 \\ 3 & 4 \end{pmatrix} \text{ and } B = \begin{pmatrix} 5 & 6 \\ 7 & 8 \end{pmatrix}$$

**A.** $\begin{pmatrix} 19 & 22 \\ 43 & 50 \end{pmatrix}$

1. Here's what the problem looks like with the numbers filled in:

$$AB = \begin{pmatrix} 1 & 2 \\ 3 & 4 \end{pmatrix}\begin{pmatrix} 5 & 6 \\ 7 & 8 \end{pmatrix}$$

2. Refer to the rule for matrix multiplication, which is

$$C_{ij} = \sum_{k=1}^{m} A_{ik} B_{kj}$$

3. When you apply the rule, you get

$$\begin{pmatrix} 1 & 2 \\ 3 & 4 \end{pmatrix}\begin{pmatrix} 5 & 6 \\ 7 & 8 \end{pmatrix} = \begin{pmatrix} 5+14 & 6+16 \\ 15+28 & 18+32 \end{pmatrix}$$

4. Wrap up the problem by adding the products together:

$$\begin{pmatrix} 5+14 & 6+16 \\ 15+28 & 18+32 \end{pmatrix} = \begin{pmatrix} 19 & 22 \\ 43 & 50 \end{pmatrix}$$

---

**3.** What's the product of **A** and **B** if

$$A = \begin{pmatrix} 1 & 1 \\ 1 & 1 \end{pmatrix} \text{ and } B = \begin{pmatrix} 2 & 2 \\ 2 & 2 \end{pmatrix}$$

*Solve It*

**4.** Find the product of **A** and **B** if

$$A = \begin{pmatrix} 1 & 2 \\ 3 & 4 \end{pmatrix} \text{ and } B = \begin{pmatrix} 5 \\ 7 \end{pmatrix}$$

*Solve It*

# Determining the Determinant

When it comes to matrices, the *determinant* is your best friend because it reduces a matrix down to a single significant number. Whether or not that number is 0 is important because that can indicate whether there's a solution to the system of equations.

The following problems give you practice finding the determinants of various matrices. Good luck!

**Q.** What's the determinant of this matrix?

$$A = \begin{pmatrix} 1 & 2 \\ 3 & 4 \end{pmatrix}$$

**A.** −2

1. Here's how the determinant is defined for a 2 × 2 matrix:

$$\det(A) = ad - cb$$

where the matrix looks like this:

$$A = \begin{pmatrix} a & b \\ c & d \end{pmatrix}$$

2. So the determinant is

$$\det(A) = (1)(4) - (3)(2)$$

which works out to be

$$\det(A) = (4) - (6) = -2$$

**5.** What's the determinant of this matrix?

$$A = \begin{pmatrix} 2 & 2 \\ 2 & 2 \end{pmatrix}$$

*Solve It*

**6.** Find the determinant of this matrix:

$$A = \begin{pmatrix} 3 & 5 \\ 7 & 9 \end{pmatrix}$$

*Solve It*

**7.** What's the determinant of this matrix?

$$A = \begin{pmatrix} 2 & 4 \\ 6 & 8 \end{pmatrix}$$

*Solve It*

**8.** Find the determinant of this matrix:

$$A = \begin{pmatrix} 9 & 8 \\ 7 & 6 \end{pmatrix}$$

*Solve It*

# More Than Just Tongue Twisters: Eigenvalues and Eigenvectors

Eigenvalues and eigenvectors are the last tools you need to solve systems of linear first order differential equations. These two items give you the ability to transform matrices into very simple forms.

Say you want to transform a matrix (such as **A**) so that when you multiply it by a vector (such as **x**), you get back **A** multiplied by some constant, which is $\lambda$.

Any values of $\lambda$ that satisfy this equation are called *eigenvalues* of the original equation. The vectors that are solutions to this equation are called *eigenvectors*.

Take a few minutes to review the following example problem; then try your hand at finding eigenvalues and eigenvectors.

**Q.** What are the eigenvalues and eigenvectors of this matrix?

$$A = \begin{pmatrix} -1 & -1 \\ 2 & -4 \end{pmatrix}$$

**A.** The eigenvalues of **A** are $\lambda_1 = -2$ and $\lambda_2 = -3$. The eigenvectors are

$$\begin{pmatrix} 1 \\ 1 \end{pmatrix}$$

and

$$\begin{pmatrix} 1 \\ 2 \end{pmatrix}$$

1. First, find $A - \lambda I$:

$$A - \lambda I = \begin{pmatrix} -1-\lambda & -1 \\ 2 & -4-\lambda \end{pmatrix}$$

2. Now find the determinant:

$$\det(A - \lambda I) = (-1 - \lambda)(-4 - \lambda) + 2$$

or

$$\det(A - \lambda I) = \lambda^2 + 5\lambda + 6$$

3. Factor this equation into

$$(\lambda + 2)(\lambda + 3)$$

So the eigenvalues of **A** are $\lambda_1 = -2$ and $\lambda_2 = -3$.

4. To find the eigenvector that corresponds to $\lambda_1$, substitute $\lambda_1$ into $A - \lambda I$:

$$A - \lambda I = \begin{pmatrix} 1 & -1 \\ 2 & -2 \end{pmatrix}$$

5. Because

$$(A - \lambda I)x = 0$$

you have

$$\begin{pmatrix} 1 & -1 \\ 2 & -2 \end{pmatrix}\begin{pmatrix} x_1 \\ x_2 \end{pmatrix} = \begin{pmatrix} 0 \\ 0 \end{pmatrix}$$

6. Every row of this matrix equation must be true, which means you can assume that $x_1 = x_2$. So, up to an arbitrary constant, the eigenvector that corresponds to $\lambda_1$ is

$$c\begin{pmatrix} 1 \\ 1 \end{pmatrix}$$

7. Drop the arbitrary constant and write the eigenvector simply as

$$\begin{pmatrix} 1 \\ 1 \end{pmatrix}$$

8. What about the eigenvector that corresponds to $\lambda_2$? Plugging $\lambda_2$ in gives you

$$\mathbf{A} - \lambda\mathbf{I} = \begin{pmatrix} 2 & -1 \\ 2 & -1 \end{pmatrix}$$

which is actually

$$\begin{pmatrix} 2 & -1 \\ 2 & -1 \end{pmatrix}\begin{pmatrix} x_1 \\ x_2 \end{pmatrix} = \begin{pmatrix} 0 \\ 0 \end{pmatrix}$$

9. So $2x_1 - x_2 = 0$ and $x_1 = x_2/2$. So, up to an arbitrary constant, the eigenvector that corresponds to $\lambda_2$ is

$$c\begin{pmatrix} 1 \\ 2 \end{pmatrix}$$

10. Good news! You can safely drop the arbitrary constant and write the eigenvector simply as

$$\begin{pmatrix} 1 \\ 2 \end{pmatrix}$$

**9.** What are the eigenvalues and eigenvectors of this matrix?

$$A = \begin{pmatrix} 2 & 1 \\ 1 & 2 \end{pmatrix}$$

*Solve It*

**10.** Find the eigenvalues and eigenvectors of this matrix:

$$A = \begin{pmatrix} 3 & -1 \\ 4 & -2 \end{pmatrix}$$

*Solve It*

# Solving Differential Equation Systems

When you have matrices and determinants, as well as eigenvalues and eigenvectors, mastered (work through the problems in the previous sections if you still need practice), you're ready to solve systems of linear first order differential equations.

Take a look at this system of homogeneous differential equations:

$$y_1' = y_1 + y_2$$
$$y_2' = 4y_1 + y_2$$

These differential equations are *linked*, which means they both contain $y_1$ and $y_2$, and therefore must be solved together. You can write them in this form:

$$\begin{pmatrix} y_1' \\ y_2' \end{pmatrix} = \begin{pmatrix} 1 & 1 \\ 4 & 1 \end{pmatrix} = \begin{pmatrix} y_1 \\ y_2 \end{pmatrix}$$

which you can then write like this:

$$\mathbf{y}' = \mathbf{Ay}$$

In this case, $\mathbf{y}'$, $\mathbf{A}$, and $\mathbf{y}$ are all matrices:

$$\mathbf{y}' = \begin{pmatrix} y_1' \\ y_2' \end{pmatrix}$$

$$\mathbf{A} = \begin{pmatrix} 1 & 1 \\ 4 & 1 \end{pmatrix}$$

$$\mathbf{y} = \begin{pmatrix} y_1 \\ y_2 \end{pmatrix}$$

If $\mathbf{A}$ is a matrix of constant coefficients, then you can assume a solution of the form

$$\mathbf{y} = \xi e^{rt}$$

No, $\xi$ isn't just a random symbol I've thrown in to see whether you're still awake. It actually stands for an eigenvector. Substituting your supposed form of the solution into the system of differential equations gives you

$$r\xi e^{rt} = \mathbf{A}\xi e^{rt}$$

Now you can subtract $\mathbf{A}\xi e^{rt}$ from both sides to get

$$(\mathbf{A} - r\mathbf{I})\xi e^{rt} = 0$$

or

$$(\mathbf{A} - r\mathbf{I})\xi = 0$$

Tada! You've just found the equation that specifies the eigenvalues and eigenvectors of matrix **A**. So the solution to this system of differential equations:

$$\mathbf{y}' = \mathbf{A}\mathbf{y}$$

is

$$\mathbf{y} = \xi e^{rt}$$

provided that $r$ is an eigenvalue of **A** and $\xi$ is the associated eigenvector.

To see the previous differential equation system worked out step by step, be sure to review the following example problem. If you're ready and raring to solve your first systems of the chapter, feel free to skip the example and dive into the practice problems instead.

**Q.** Find the solution to this system of differential equations:

$$y_1' = y_1 + y_2$$
$$y_2' = 4y_1 + y_2$$

**A.** $y_1 = c_1 e^{3t} - c_2 e^{-t}$ and $y_2 = 2c_1 e^{3t} + 2c_2 e^{-t}$

1. Write this problem as

$$\begin{pmatrix} y_1' \\ y_2' \end{pmatrix} = \begin{pmatrix} 1 & 1 \\ 4 & 1 \end{pmatrix} = \begin{pmatrix} y_1 \\ y_2 \end{pmatrix}$$

2. Because this system has constant coefficients, try a solution of the form

$$\mathbf{y} = \xi e^{rt}$$

3. Substituting your attempted solution into the system gives you

$$\begin{pmatrix} r\xi_1 e^{rt} \\ r\xi_2 e^{rt} \end{pmatrix} = \begin{pmatrix} 1 & 1 \\ 4 & 1 \end{pmatrix}\begin{pmatrix} \xi_1 e^{rt} \\ \xi_2 e^{rt} \end{pmatrix}$$

which you can rewrite as

$$\begin{pmatrix} 0 \\ 0 \end{pmatrix} = \begin{pmatrix} 1-r & 1 \\ 4 & 1-r \end{pmatrix}\begin{pmatrix} \xi_1 e^{rt} \\ \xi_2 e^{rt} \end{pmatrix}$$

4. Divide by $e^{rt}$ to get

$$\begin{pmatrix} 0 \\ 0 \end{pmatrix} = \begin{pmatrix} 1-r & 1 \\ 4 & 1-r \end{pmatrix}\begin{pmatrix} \xi_1 \\ \xi_2 \end{pmatrix}$$

5. This system of linear equations has a (nontrivial) solution only if the determinant of the $2 \times 2$ matrix is 0, so

$$\det\begin{pmatrix} 1-r & 1 \\ 4 & 1-r \end{pmatrix} = 0$$

6. Expanding the determinant gives you

$$(1-r)(1-r) - 4 = 0$$

which becomes

$$r^2 - 2r + 1 - 4 = 0$$

or

$$r^2 - 2r - 3 = 0$$

7. Factor the characteristic equation as follows:

$$(r-3)(r+1) = 0$$

to reveal that the eigenvalues of the matrix are

$$r_1 = 3 \text{ and } r_2 = -1$$

8. Now you need to find the two eigenvectors. Start that process by taking the first eigenvalue, $r_1 = 3$, and plugging it in:

$$\begin{pmatrix} 0 \\ 0 \end{pmatrix} = \begin{pmatrix} -2 & 1 \\ 4 & -2 \end{pmatrix}\begin{pmatrix} \xi_1 \\ \xi_2 \end{pmatrix}$$

Doing so gives you

$$-2\xi_1 + \xi_2 = 0$$

and

$$4\xi_1 - 2\xi_2 = 0$$

9. These equations are the same up to a factor of –1, so

$$2\xi_1 = \xi_2$$

Therefore, the first eigenvector (up to an arbitrary constant, of course) is

$$\begin{pmatrix} \xi_1 \\ \xi_2 \end{pmatrix} = \begin{pmatrix} 1 \\ 2 \end{pmatrix}$$

10. Now on to the second eigenvector. This one corresponds to the eigenvalue $r_2 = -1$:

$$\begin{pmatrix} 0 \\ 0 \end{pmatrix} = \begin{pmatrix} 1-r & 1 \\ 4 & 1-r \end{pmatrix} \begin{pmatrix} \xi_1 \\ \xi_2 \end{pmatrix}$$

11. Plug $r_2 = -1$ into the preceding matrix to get

$$\begin{pmatrix} 0 \\ 0 \end{pmatrix} = \begin{pmatrix} 2 & 1 \\ 4 & 2 \end{pmatrix} \begin{pmatrix} \xi_1 \\ \xi_2 \end{pmatrix}$$

which gives you

$$2\xi_1 + \xi_2 = 0$$

and

$$4\xi_1 + 2\xi_2 = 0$$

12. These two equations offer you the same information: the fact that $2\xi_1 = -\xi_2$. So the second eigenvector becomes

$$\begin{pmatrix} \xi_1 \\ \xi_2 \end{pmatrix} = \begin{pmatrix} -1 \\ 2 \end{pmatrix}$$

13. Therefore, the first solution to the system is

$$\begin{pmatrix} 1 \\ 2 \end{pmatrix} e^{3t}$$

and the second solution is

$$\begin{pmatrix} -1 \\ 2 \end{pmatrix} e^{-t}$$

14. As you can see, the general solution is a linear combination of the two solutions:

$$\mathbf{y} = c_1 \begin{pmatrix} 1 \\ 2 \end{pmatrix} e^{3t} + c_2 \begin{pmatrix} -1 \\ 2 \end{pmatrix} e^{-t}$$

which can also be written as

$$\begin{pmatrix} y_1 \\ y_2 \end{pmatrix} = c_1 \begin{pmatrix} 1 \\ 2 \end{pmatrix} e^{3t} + c_2 \begin{pmatrix} -1 \\ 2 \end{pmatrix} e^{-t}$$

15. So the solution to the system of differential equations is

$$y_1 = c_1 e^{3t} - c_2 e^{-t} \text{ and}$$
$$y_2 = 2c_1 e^{3t} + 2c_2 e^{-t}$$

**11.** Find the solution to this system of differential equations:

$$y_1' = -y_1 - y_2$$
$$y_2' = 2y_1 - 4y_2$$

*Solve It*

**12.** What's the solution to this system of differential equations?

$$y_1' = 3y_1 + 2y_2$$
$$y_2' = 4y_1 + y_2$$

*Solve It*

# Answers to Systems of Linear First Order Differential Equation Problems

Following are the answers to the practice questions presented throughout this chapter. Each one is worked out step by step so that if you messed one up along the way, you can more easily see where you took a wrong turn.

**1** Find the sum of A + B if A and B are

$$A = \begin{pmatrix} 4 & 3 \\ 2 & 1 \end{pmatrix} \text{ and } B = \begin{pmatrix} 1 & 2 \\ 3 & 4 \end{pmatrix}$$

Solution: $A + B = \begin{pmatrix} 5 & 5 \\ 5 & 5 \end{pmatrix}$

1. A + B is

$$\begin{pmatrix} 4 & 3 \\ 2 & 1 \end{pmatrix} + \begin{pmatrix} 1 & 2 \\ 3 & 4 \end{pmatrix}$$

2. By adding element to element, you get

$$\begin{pmatrix} 4 & 3 \\ 2 & 1 \end{pmatrix} + \begin{pmatrix} 1 & 2 \\ 3 & 4 \end{pmatrix} = \begin{pmatrix} 5 & 5 \\ 5 & 5 \end{pmatrix}$$

**2** What's A + B if A and B are

$$A = \begin{pmatrix} 9 & 9 \\ 9 & 9 \end{pmatrix} \text{ and } B = \begin{pmatrix} 1 & 2 \\ 3 & 4 \end{pmatrix}$$

Solution: $A + B = \begin{pmatrix} 10 & 11 \\ 12 & 13 \end{pmatrix}$

1. Here's what **A** + **B** looks like written out fully:

$$\begin{pmatrix} 9 & 9 \\ 9 & 9 \end{pmatrix} + \begin{pmatrix} 1 & 2 \\ 3 & 4 \end{pmatrix}$$

2. Add the corresponding elements to each other to find the solution:

$$\begin{pmatrix} 9 & 9 \\ 9 & 9 \end{pmatrix} + \begin{pmatrix} 1 & 2 \\ 3 & 4 \end{pmatrix} = \begin{pmatrix} 10 & 11 \\ 12 & 13 \end{pmatrix}$$

**3** What's the product of A and B if

$$A = \begin{pmatrix} 1 & 1 \\ 1 & 1 \end{pmatrix} \text{ and } B = \begin{pmatrix} 2 & 2 \\ 2 & 2 \end{pmatrix}$$

Solution: $\begin{pmatrix} 4 & 4 \\ 4 & 4 \end{pmatrix}$

1. Here's what the problem looks like with the numbers filled in:

$$AB = \begin{pmatrix} 1 & 1 \\ 1 & 1 \end{pmatrix} \begin{pmatrix} 2 & 2 \\ 2 & 2 \end{pmatrix}$$

2. Refer to the rule for matrix multiplication, which is

$$C_{ij} = \sum_{k=1}^{m} A_{ik} B_{kj}$$

3. When you apply the rule, you get

$$\begin{pmatrix} 1 & 1 \\ 1 & 1 \end{pmatrix} \begin{pmatrix} 2 & 2 \\ 2 & 2 \end{pmatrix} = \begin{pmatrix} 2+2 & 2+2 \\ 2+2 & 2+2 \end{pmatrix}$$

4. Wrap up the problem by adding the products together:

$$\begin{pmatrix} 2+2 & 2+2 \\ 2+2 & 2+2 \end{pmatrix} = \begin{pmatrix} 4 & 4 \\ 4 & 4 \end{pmatrix}$$

**4** **Find the product of A and B if**

$$A = \begin{pmatrix} 1 & 2 \\ 3 & 4 \end{pmatrix} \text{ and } B = \begin{pmatrix} 5 \\ 7 \end{pmatrix}$$

**Solution:** $\begin{pmatrix} 19 \\ 43 \end{pmatrix}$

1. When you fill in the numbers, the problem in Question 4 looks like this:

$$AB = \begin{pmatrix} 1 & 2 \\ 3 & 4 \end{pmatrix} \begin{pmatrix} 5 \\ 7 \end{pmatrix}$$

2. Recall the rule for matrix multiplication:

$$C_{ij} = \sum_{k=1}^{m} A_{ik} B_{kj}$$

3. Applying the rule gives you

$$\begin{pmatrix} 1 & 2 \\ 3 & 4 \end{pmatrix} \begin{pmatrix} 5 \\ 7 \end{pmatrix} = \begin{pmatrix} 5+14 \\ 15+28 \end{pmatrix}$$

4. So your final solution is as follows

$$\begin{pmatrix} 5+14 \\ 15+28 \end{pmatrix} = \begin{pmatrix} 19 \\ 43 \end{pmatrix}$$

**5** **What's the determinant of this matrix?**

$$A = \begin{pmatrix} 2 & 2 \\ 2 & 2 \end{pmatrix}$$

**Solution: 0**

1. Here's how the determinant is defined for a 2 × 2 matrix:

$$\det(A) = ad - cb$$

where the matrix looks like this:

$$A = \begin{pmatrix} a & b \\ c & d \end{pmatrix}$$

2. So the determinant is

$$\det(A) = (2)(2) - (2)(2)$$

which works out to be

$$\det(A) = (4) - (4) = 0$$

**6** **Find the determinant of this matrix:**

$$A = \begin{pmatrix} 3 & 5 \\ 7 & 9 \end{pmatrix}$$

**Solution: –8**

1. Because the determinant for a 2 × 2 matrix is defined as

$$\det(A) = ad - cb$$

where the matrix looks like this:

$$A = \begin{pmatrix} a & b \\ c & d \end{pmatrix}$$

the determinant is

$$\det(A) = (3)(9) - (7)(5)$$

2. That makes the solution

$$\det(A) = (27) - (35) = -8$$

**7** **What's the determinant of this matrix?**

$$A = \begin{pmatrix} 2 & 4 \\ 6 & 8 \end{pmatrix}$$

**Solution: –8**

1. Here's how the determinant is defined for a 2 × 2 matrix:

$$\det(A) = ad - cb$$

where the matrix looks like this:

$$A = \begin{pmatrix} a & b \\ c & d \end{pmatrix}$$

2. So the determinant is

$$\det(A) = (2)(8) - (4)(6)$$

which works out to be

$$\det(A) = (16) - (24) = -8$$

**8** **Find the determinant of this matrix:**

$$A = \begin{pmatrix} 9 & 8 \\ 7 & 6 \end{pmatrix}$$

**Solution: –2**

1. Because the determinant for a 2 × 2 matrix is defined as

   det(**A**) = $ad - cb$

   where the matrix looks like this:

   $$A = \begin{pmatrix} a & b \\ c & d \end{pmatrix}$$

   the determinant is

   det(**A**) = (9)(6) – (8)(7)

2. That makes the solution

   det(**A**) = (54) – (56) = –2

9  **What are the eigenvalues and eigenvectors of this matrix?**

   $$A = \begin{pmatrix} 2 & 1 \\ 1 & 2 \end{pmatrix}$$

**Solution: The eigenvalues of A are $\lambda_1 = 1$ and $\lambda_2 = 3$. The eigenvectors are**

$$\begin{pmatrix} 1 \\ -1 \end{pmatrix}$$

**and**

$$\begin{pmatrix} 1 \\ 1 \end{pmatrix}$$

1. First, find **A** – λ**I**:

   $$A - \lambda I = \begin{pmatrix} 2-\lambda & 1 \\ 1 & 2-\lambda \end{pmatrix}$$

2. Now find the determinant, which is

   det(**A** – λ**I**) = $(2 - \lambda)(2 - \lambda) - 1$

   or

   det(**A** – λ**I**) = $\lambda^2 - 4\lambda + 3$

3. Factor this equation into

   $(\lambda - 1)(\lambda - 3)$

   So the eigenvalues of **A** are $\lambda_1 = 1$ and $\lambda_2 = 3$.

4. To find the eigenvector that corresponds to $\lambda_1$, substitute $\lambda_1$ into **A** – λ**I**:

   $$A - \lambda I = \begin{pmatrix} 1 & 1 \\ 1 & 1 \end{pmatrix}$$

5. Because

   (**A** –λ**I**)**x** = 0

   you have

   $$\begin{pmatrix} 1 & 1 \\ 1 & 1 \end{pmatrix}\begin{pmatrix} x_1 \\ x_2 \end{pmatrix} = \begin{pmatrix} 0 \\ 0 \end{pmatrix}$$

6. Every row of this matrix equation must be true, which means you can assume that $x_1 = -x_2$. So, up
an arbitrary constant, the eigenvector that corresponds to $\lambda_1$ is

$$c\begin{pmatrix} 1 \\ -1 \end{pmatrix}$$

7. Drop the arbitrary constant and write the eigenvector simply as

$$\begin{pmatrix} 1 \\ -1 \end{pmatrix}$$

8. What about the eigenvector that corresponds to $\lambda_2$? Plugging $\lambda_2$ in gives you

$$\mathbf{A} - \lambda \mathbf{I} = \begin{pmatrix} -1 & 1 \\ 1 & -1 \end{pmatrix}$$

which is actually

$$\begin{pmatrix} -1 & 1 \\ 1 & -1 \end{pmatrix}\begin{pmatrix} x_1 \\ x_2 \end{pmatrix} = \begin{pmatrix} 0 \\ 0 \end{pmatrix}$$

9. So $x_1 = 0$, which means that, up to an arbitrary constant, the eigenvector that corresponds to $\lambda_2$ is

$$c\begin{pmatrix} 1 \\ 1 \end{pmatrix}$$

10. Good news! You can drop the arbitrary constant and just write the eigenvector simply as

$$\begin{pmatrix} 1 \\ 1 \end{pmatrix}$$

**10** **Find the eigenvalues and eigenvectors of this matrix:**

$$\mathbf{A} = \begin{pmatrix} 3 & -1 \\ 4 & -2 \end{pmatrix}$$

**Solution: The eigenvalues of A are $\lambda_1 = 2$ and $\lambda_2 = -1$. The eigenvectors are**

$$\begin{pmatrix} 1 \\ 1 \end{pmatrix}$$

**and**

$$\begin{pmatrix} 1 \\ 4 \end{pmatrix}$$

1. Start off by finding $\mathbf{A} - \lambda \mathbf{I}$:

$$\mathbf{A} - \lambda \mathbf{I} = \begin{pmatrix} 3 - \lambda & -1 \\ 4 & -2 - \lambda \end{pmatrix}$$

2. Then find the determinant:

$$\det(\mathbf{A} - \lambda \mathbf{I}) = (3 - \lambda)(-2 - \lambda) + 4$$

which equals

$$\lambda^2 - \lambda - 2$$

3. Factor this equation as follows:

$$(\lambda + 1)(\lambda - 2)$$

to reveal that the eigenvalues of **A** are $\lambda_1 = 2$ and $\lambda_2 = -1$.

4. Now it's time to find the eigenvectors. Substitute $\lambda_1$ into $\mathbf{A} - \lambda\mathbf{I}$ to find the eigenvector corresponding to $\lambda_1$:

$$\mathbf{A} - \lambda\mathbf{I} = \begin{pmatrix} 1 & -1 \\ 4 & -4 \end{pmatrix}$$

5. You already know that

$$(\mathbf{A} - \lambda\mathbf{I})\mathbf{x} = 0$$

so

$$\begin{pmatrix} 1 & -1 \\ 4 & -4 \end{pmatrix} \begin{pmatrix} x_1 \\ x_2 \end{pmatrix} = \begin{pmatrix} 0 \\ 0 \end{pmatrix}$$

6. Because every row of this matrix equation must be true, go ahead and assume that $x_1 = -x_2$. So, up to an arbitrary constant, the eigenvector corresponding to $\lambda_1$ is

$$c\begin{pmatrix} 1 \\ 1 \end{pmatrix}$$

7. Drop the arbitrary constant and write the eigenvector as follows:

$$\begin{pmatrix} 1 \\ 1 \end{pmatrix}$$

8. Great. Now plug $\lambda_2$ in to get

$$\mathbf{A} - \lambda\mathbf{I} = \begin{pmatrix} 4 & -1 \\ 4 & -1 \end{pmatrix}$$

so

$$\begin{pmatrix} 4 & -1 \\ 4 & -1 \end{pmatrix} \begin{pmatrix} x_1 \\ x_2 \end{pmatrix} = \begin{pmatrix} 0 \\ 0 \end{pmatrix}$$

9. Looks like $4x_1 = x_2$, which means that, up to an arbitrary constant, the eigenvector corresponding to $\lambda_2$ is

$$c\begin{pmatrix} 1 \\ 4 \end{pmatrix}$$

10. For simplicity's sake, drop the arbitrary constant:

$$\begin{pmatrix} 1 \\ 4 \end{pmatrix}$$

**11** **Find the solution to this system of differential equations:**

$$y_1' = -y_1 - y_2$$
$$y_2' = 2y_1 - 4y_2$$

**Solution:** $y_1 = c_1 e^{-2t} + c_2 e^{-3t}$ and $y_2 = c_1 e^{-2t} + 2c_2 e^{-3t}$

1. Write this problem as

$$\begin{pmatrix} y_1' \\ y_2' \end{pmatrix} = \begin{pmatrix} -1 & -1 \\ 2 & -4 \end{pmatrix} \begin{pmatrix} y_1 \\ y_2 \end{pmatrix}$$

2. Because this system has constant coefficients, try a solution of the form

$$\mathbf{y} = \xi e^{rt}$$

3. Substituting your attempted solution into the system gives you

$$\begin{pmatrix} r\xi_1 e^{rt} \\ r\xi_2 e^{rt} \end{pmatrix} = \begin{pmatrix} -1 & -1 \\ 2 & -4 \end{pmatrix} \begin{pmatrix} \xi_1 e^{rt} \\ \xi_2 e^{rt} \end{pmatrix}$$

which you can rewrite as

$$\begin{pmatrix} 0 \\ 0 \end{pmatrix} = \begin{pmatrix} -1-r & -1 \\ 2 & -4-r \end{pmatrix} \begin{pmatrix} \xi_1 e^{rt} \\ \xi_2 e^{rt} \end{pmatrix}$$

4. Divide by $e^{rt}$ to get

$$\begin{pmatrix} 0 \\ 0 \end{pmatrix} = \begin{pmatrix} -1-r & -1 \\ 2 & -4-r \end{pmatrix} \begin{pmatrix} \xi_1 \\ \xi_2 \end{pmatrix}$$

5. This system of linear equations has a (nontrivial) solution only if the determinant of the $2 \times 2$ matrix is 0, so

$$\det \begin{pmatrix} -1-r & -1 \\ 2 & -4-r \end{pmatrix} = 0$$

6. Expanding the determinant gives you

$$(-1-r)(-4-r) + 2 = 0$$

which becomes

$$(1+r)(4+r) + 2 = 0$$

or

$$r^2 + 5r + 4 + 2 = 0$$

which is also

$$r^2 + 5r + 6 = 0$$

7. Factor the characteristic equation as follows:

$$(r+2)(r+3) = 0$$

to reveal that the eigenvalues of the matrix are

$$r_1 = -2 \text{ and } r_2 = -3$$

8. Now you need to find the two eigenvectors. Start that process by taking the first eigenvalue, $r_1 = -2$, and plugging it in:

$$\begin{pmatrix} 0 \\ 0 \end{pmatrix} = \begin{pmatrix} 1 & -1 \\ 2 & -2 \end{pmatrix} \begin{pmatrix} \xi_1 \\ \xi_2 \end{pmatrix}$$

Doing so gives you

$$\xi_1 - \xi_2 = 0$$

and

$$2\xi_1 - 2\xi_2 = 0$$

9. These equations are the same up to a factor of $-1$, so

$$\xi_1 = \xi_2$$

Therefore, the first eigenvector (up to an arbitrary constant, of course) is

$$\begin{pmatrix} \xi_1 \\ \xi_2 \end{pmatrix} = \begin{pmatrix} 1 \\ 1 \end{pmatrix}$$

10. Now on to the second eigenvector. This one corresponds to the eigenvalue $r_2 = -3$:

$$\begin{pmatrix} 0 \\ 0 \end{pmatrix} = \begin{pmatrix} -1-r & -1 \\ 2 & -4-r \end{pmatrix}\begin{pmatrix} \xi_1 \\ \xi_2 \end{pmatrix}$$

11. Plug $r_2 = -3$ into the preceding matrix to get

$$\begin{pmatrix} 0 \\ 0 \end{pmatrix} = \begin{pmatrix} 2 & -1 \\ 2 & -1 \end{pmatrix}\begin{pmatrix} \xi_1 \\ \xi_2 \end{pmatrix}$$

which gives you

$$2\xi_1 - \xi_2 = 0$$

and

$$2\xi_1 - \xi_2 = 0$$

12. These two equations offer you the same info: the fact that $2\xi_1 = -\xi_2$. So the second eigenvector becomes

$$\begin{pmatrix} \xi_1 \\ \xi_2 \end{pmatrix} = \begin{pmatrix} 1 \\ 2 \end{pmatrix}$$

13. Therefore, the first solution to the system is

$$\begin{pmatrix} 1 \\ 1 \end{pmatrix} e^{-2t}$$

and the second solution is

$$\begin{pmatrix} 1 \\ 2 \end{pmatrix} e^{-3t}$$

14. As you can see, the general solution is a linear combination of the two solutions:

$$\mathbf{y} = c_1 \begin{pmatrix} 1 \\ 1 \end{pmatrix} e^{-2t} + c_2 \begin{pmatrix} 1 \\ 2 \end{pmatrix} e^{-3t}$$

which can also be written as

$$\begin{pmatrix} y_1 \\ y_2 \end{pmatrix} = c_1 \begin{pmatrix} 1 \\ 2 \end{pmatrix} e^{3t} + c_2 \begin{pmatrix} -1 \\ 2 \end{pmatrix} e^{-t}$$

15. So the solution to the system of differential equations is

$$y_1 = c_1 e^{-2t} + c_2 e^{-3t} \text{ and } y_2 = c_1 e^{-2t} + 2c_2 e^{-3t}$$

*12*  **What's the solution to this system of differential equations?**

$$y_1' = 3y_1 + 2y_2$$
$$y_2' = 4y_1 + y_2$$

**Solution: $y_1 = c_1 e^{-t} + c_2 e^{5t}$ and $y_2 = -2c_1 e^{-t} + c_2 e^{5t}$**

1. Rewrite the original problem as follows:

$$\begin{pmatrix} y_1' \\ y_2' \end{pmatrix} = \begin{pmatrix} 3 & 2 \\ 4 & 1 \end{pmatrix} \begin{pmatrix} y_1 \\ y_2 \end{pmatrix}$$

2. This system has constant coefficients, so go ahead and try a solution of this form:

$$\mathbf{y} = \xi e^{rt}$$

3. Then substitute your attempted solution into the system:

$$\begin{pmatrix} r\xi_1 e^{rt} \\ r\xi_2 e^{rt} \end{pmatrix} = \begin{pmatrix} 3 & 2 \\ 4 & 1 \end{pmatrix} \begin{pmatrix} \xi_1 e^{rt} \\ \xi_2 e^{rt} \end{pmatrix}$$

You can rewrite that result as

$$\begin{pmatrix} 0 \\ 0 \end{pmatrix} = \begin{pmatrix} 3-r & 2 \\ 4 & 1-r \end{pmatrix} \begin{pmatrix} \xi_1 e^{rt} \\ \xi_2 e^{rt} \end{pmatrix}$$

4. Dividing by $e^{rt}$ gives you

$$\begin{pmatrix} 0 \\ 0 \end{pmatrix} = \begin{pmatrix} 3-r & 2 \\ 4 & 1-r \end{pmatrix} \begin{pmatrix} \xi_1 \\ \xi_2 \end{pmatrix}$$

5. This system of linear equations has a (nontrivial) solution only if the determinant of the $2 \times 2$ matrix is 0, so

$$\det \begin{pmatrix} 3-r & 2 \\ 4 & 1-r \end{pmatrix} = 0$$

6. Expand the determinant to get

$$(3-r)(1-r) - 8 = 0$$

which becomes

$$r^2 - 4r + 3 - 8 = 0$$

or

$$r^2 - 4r - 5 = 0$$

7. Factor the characteristic equation into

$$(r+1)(r-5) = 0$$

8. Now you can see that the eigenvalues of the matrix are

$$r_1 = -1 \text{ and } r_2 = 5$$

9. Time to determine the two eigenvectors. First, take $r_1 = -1$ (the first eigenvalue) and plug it in to get

$$\begin{pmatrix} 0 \\ 0 \end{pmatrix} = \begin{pmatrix} 4 & 2 \\ 4 & 2 \end{pmatrix} \begin{pmatrix} \xi_1 \\ \xi_2 \end{pmatrix}$$

or

$$4\xi_1 + 2\xi_2 = 0$$

and

$$4\xi_1 + 2\xi_2 = 0$$

10. These equations give you the same info, which is that

$$2\xi_1 = -\xi_2$$

So, up to an arbitrary constant, the first eigenvector is

$$\begin{pmatrix} \xi_1 \\ \xi_2 \end{pmatrix} = \begin{pmatrix} 1 \\ -2 \end{pmatrix}$$

11. How about the second eigenvector? Well, you know that it corresponds to the eigenvalue $r_2 = 5$:

$$\begin{pmatrix} 0 \\ 0 \end{pmatrix} = \begin{pmatrix} 3-r & 2 \\ 4 & 1-r \end{pmatrix} \begin{pmatrix} \xi_1 \\ \xi_2 \end{pmatrix}$$

12. Plugging $r_2 = 5$ into the preceding equation gives you

$$\begin{pmatrix} 0 \\ 0 \end{pmatrix} = \begin{pmatrix} -2 & 2 \\ 4 & -4 \end{pmatrix} \begin{pmatrix} \xi_1 \\ \xi_2 \end{pmatrix}$$

so you wind up with

$$-2\xi_1 + 2\xi_2 = 0$$

and

$$4\xi_1 - 4\xi_2 = 0$$

13. Surprise, surprise: These equations give you the same info, which is that $\xi_1 = \xi_2$. The second eigenvector therefore becomes

$$\begin{pmatrix} \xi_1 \\ \xi_2 \end{pmatrix} = \begin{pmatrix} 1 \\ 1 \end{pmatrix}$$

14. Looks like the first solution to the system is

$$\begin{pmatrix} 1 \\ -2 \end{pmatrix} e^{-t}$$

and the second solution is

$$\begin{pmatrix} 1 \\ 1 \end{pmatrix} e^{5t}$$

15. The general solution you're going for is a linear combination of the two solutions, which looks like this:

$$\mathbf{y} = c_1 \begin{pmatrix} 1 \\ -2 \end{pmatrix} e^{-t} + c_2 \begin{pmatrix} 1 \\ 1 \end{pmatrix} e^{5t}$$

which can also be written as

$$\begin{pmatrix} y_1 \\ y_2 \end{pmatrix} = c_1 \begin{pmatrix} 1 \\ -2 \end{pmatrix} e^{-t} + c_2 \begin{pmatrix} 1 \\ 1 \end{pmatrix} e^{5t}$$

16. Thus, the solution to the system of differential equations is

$$y_1 = c_1 e^{-t} + c_2 e^{5t} \text{ and } y_2 = -2c_1 e^{-t} + c_2 e^{5t}$$

# Part IV
# The Part of Tens

The 5th Wave                    By Rich Tennant

I'm mathematically dyslexic. But it's not that unusual — 100 out of every 15 people are.

## In this part . . .

*I*f you're a fan of top ten lists (or even if you just want a break from tackling all the great practice problems throughout this workbook), then this is the part for you. First, I give you a tour of the ten common ways of solving differential equations, complete with online resources. And because differential equations don't exist in a vacuum (believe it or not, they're meant to solve real-world problems), I also show you ten real-world applications for them.

# Chapter 12

# Ten Common Ways of Solving Differential Equations

## In This Chapter

▶ Surveying the different types of differential equations

▶ Taking stock of your solution-technique options

Tackling differential equations effectively — and with the least amount of frustration — means knowing what type of equation you're dealing with and having a solution technique in mind. For example, are you looking at a homogeneous or separable differential equation, and can you solve it with either a series or numerical solution?

Consider this chapter your overview of the ten common ways of solving differential equations and where to find online help.

## Looking at Linear Equations

*Linear* differential equations exclusively involve linear terms (meaning terms to the first power) of $y$, $y'$, $y''$, $y'''$, and so on. An equation that looks like this is considered linear:

$$y'' + 3y' + 6y - 4 = 0$$

For a great explanation of linear first order differential equations and how to solve them, visit www.sosmath.com/diffeq/diffeq.html, look for the First Order Differential Equations bullet, and click the Linear Equations link. Then flip to Chapter 1 for some practice solving linear first order differential equations.

## Scoping Out Separable Equations

*Separable* differential equations can be written so that all $x$ terms appear on one side of the equal sign and all other terms appear on the opposite side. Here's an example:

$$\frac{dy}{dx} = x^4 - x^2$$

This differential equation can be separated as

$$dy = x^4\, dx - x^2\, dx$$

For additional help spotting and understanding separable differential equations, head to www.sosmath.com/diffeq/diffeq.html, find the First Order Differential Equations bullet, and click the Separable Equations link. Or if you're feeling up to it, head to Chapter 2 to start solving separable differential equations.

# Applying the Method of Undetermined Coefficients

When your differential equation has constant coefficients, like this one does:

$$y'' + 9y' + 8y - 4 = 0$$

then you should try the method of undetermined coefficients to solve it. Flip to Chapter 5 for a demonstration of this technique; then check out this additional resource: tutorial.math.lamar.edu/Classes/DE/UndeterminedCoefficients.aspx.

# Honing in on Homogeneous Equations

In a *homogeneous* differential equation, all the terms involve *y,* as you can see in this example:

$$y'' - 7y' + 12y = 0$$

You typically write homogeneous differential equations by setting the right side of the equation equal to 0. For some online help with recognizing and solving homogeneous equations, check out en.wikipedia.org/wiki/Homogeneous_differential_equation.

# Examining Exact Equations

If you encounter a differential equation that can be written in this form:

$$M(x, y)\, dx + N(x, y)\, dy = 0$$

then the equation can be called exact if

$$\frac{dM\,(x,\ y)}{dy} = \frac{dN\,(x,\ y)}{dx}$$

Want to check out exact differential equations on the Web? Visit `www.sosmath.com/diffeq/diffeq.html`, find the First Order Differential Equations bullet, and click the Exact and Non-Exact Equations link. For some practice solving exact differential equations, flip to Chapter 3.

# Finding Solutions with the Help of Integrating Factors

Whenever you see a differential equation like the following:

$$M(x, y)\, dx + N(x, y)\, dy = 0$$

but it isn't exact — that is, this statement is true:

$$\frac{dM\,(x,\,y)}{dy} \neq \frac{dN\,(x,\,y)}{dx}$$

then you can try to find an integrating factor — $\mu(x, y)$ — such that the differential equation changes form to

$$\mu(x, y)M(x, y)\, dx + \mu(x, y)N(x, y)\, dy = 0$$

and becomes exact.

If you want a Web-based review on finding integrating factors, check out `www.sosmath.com/diffeq/diffeq.html`, look for the First Order Differential Equations bullet, and click the Integrating Factor technique bullet.

# Getting Serious Answers with Series Solutions

Is a tough differential equation like this one bogging you down?

$$y'' + xy' + 2y = 0$$

Then try solving it with a series solution, which allows you to assume that $y$ can be expanded in a power series like this:

$$y = \sum_{n=0}^{\infty} a_n x^n$$

For practice solving differential equations using series solutions, flip to Chapter 8. You can also check out this great online resource: `tutorial.math.lamar.edu/Classes/DE/SeriesSolutions.aspx`.

# Turning to Laplace Transforms for Solutions

Laplace transforms offer you a powerful tool for solving differential equations like the following:

$$y'' + 5y' + 6y = 0$$

Take the Laplace transform of this differential equation and apply any initial conditions to get, for example,

$$\mathcal{L}\{y\} = \frac{1}{(s+2)} + \frac{1}{(s+3)}$$

Then take the inverse Laplace transform to get the solution to the differential equation:

$$y = e^{-2x} + e^{-3x}$$

Check out Chapter 10 for some practice using Laplace transforms, or head to `tutorial.math.lamar.edu/Classes/DE/IVPWithLaplace.aspx` for some additional refresher.

# Determining whether a Solution Exists

Sometimes a differential equation may not have a solution. Fortunately, a number of theorems are available to help you determine whether that's the case. For more on the existence and uniqueness of solutions, head to `www.sosmath.com/diffeq/diffeq.html`, find the First Order Differential Equations bullet, and click the Existence and Uniqueness of Solutions link.

# Solving Equations with Computer-Based Numerical Methods

Computer-based numerical methods are always an option when you're faced with crazy, complex differential equations, such as

$$\sin (y)y^{(4)} - 93 \cos (x)y''' + 3.7y' + 6y^6 - 4e^y = \sin (x) \cosh (x)$$

Several popular mathematical techniques, such as Euler's method and the Runge-Kutta method, can be readily translated into computer code. Read all about these methods in *Differential Equations For Dummies* or at `www.efunda.com/math/num_ode/num_ode.cfm`.

# Chapter 13

# Ten Real-World Applications of Differential Equations

***In This Chapter***
▶ Using differential equations to determine growth or decay
▶ Discovering motion-related facts with the help of differential equations

*E*ver find yourself wondering what's so good about knowing how to solve differential equations — besides being able to complete problem sets? Well, differential equations are all about letting you model the real world mathematically, and in this chapter, you get a list of the ten best real-world uses for differential equations, along with Web sites that carry out these uses. (This chapter is just the tip of the iceberg, of course; an infinite number of real-world applications exist for differential equations.)

## Calculating Population Growth

The rate of population change is proportional to the current size of the population, as shown in the following differential equation:

$$\frac{dP}{dt} = kP$$

Note that *P* is the population, and *k* is a constant. For a good look at the solutions of this equation, head to www.analyzemath.com/calculus/Differential_Equations and click the applications.html link. Population growth is addressed in Application 1.

## Determining Fluid Flow

Fluid in a pipe moves faster near the center of the pipe and slower near the pipe's walls. You can find the velocity as a function of *r*, the radius from the center of the pipe, with this equation:

$$\frac{d^2v}{dr^2} + \frac{1}{r}\frac{dv}{dr} = \frac{-1}{\eta}\frac{\Delta P}{\Delta x}$$

where $v$ (the velocity of the fluid) is a function of $r$, $\eta$ is the fluid's viscosity, and $\Delta P/\Delta x$ is the pressure gradient. See more at `hyperphysics.phy-astr.gsu.edu/hbase/pfric2.html#vpro2`.

# Mixing Fluids

When you mix fluids in tanks, you can relate the mass of a certain substance with the flow rate and concentration as follows:

$$\frac{dm}{dt} = -qC$$

In this equation, $m$ is mass, $t$ is time, $q$ is the flow rate, and $C$ is concentration. To get all the details on mixing fluids in tanks, visit `www.tmt.ugal.ro/crios/Support/ANPT/Curs/deqn/a1/stanks2/stanks2.html`.

# Finding Out Facts about Falling Objects

A basic use of differential equations is finding out information about falling objects. This equation tells you the object's acceleration:

$$a(t) = \frac{dv}{dt}$$

whereas this equation tells you the object's velocity:

$$v(t) = \frac{dy}{dt}$$

For more details, go to `www.analyzemath.com/calculus/Differential_Equations`, click the applications.html link, and scroll to Application 3.

# Calculating Trajectories

You can describe the *trajectories* (paths) of objects with this differential equation:

$$2x + 2y\,\frac{dy}{dx} = C$$

where $x$ and $y$ are the standard coordinates of the object and $C$ is a constant. For more on using differential equations to describe the trajectories of objects, take a look at `www.sosmath.com/diffeq/diffeq.html`, find the First Order Differential Equations bullet, and click the Orthogonal Trajectories link.

# Analyzing the Motion of Pendulums

Need to figure out the specifics of a pendulum's motion? Just bust out the following differential equation:

$$\frac{d^2\theta}{dt^2} + \frac{g\theta}{L} = 0$$

where $\theta$ is the angle of the pendulum at any given moment, $L$ is the length of the pendulum, $t$ is time, and $g$ is the acceleration due to gravity. For a solution that describes the motion of a pendulum, check out hyperphysics.phy-astr.gsu.edu/hbase/pend.html#c5.

# Applying Newton's Law of Cooling

Newton's law of cooling says that the rate of temperature change of an object is proportional to the temperature difference of that object from the environment:

$$\frac{dT}{dt} = -k\left(T - T_e\right)$$

In this equation, $T$ is the temperature of an object, $t$ is time, $k$ is a constant, and $T_e$ is the temperature of the environment. You can see this differential equation solved by going to www.analyzemath.com/calculus/Differential_Equations, clicking the applications.html link, and scrolling to Application 4.

# Determining Radioactive Decay

Atoms in radioactive materials decay at a certain rate, which is given by this differential equation:

$$\frac{dN}{N} = -\lambda \ dt$$

where $N$ is the number of atoms of the radioactive material, $\lambda$ is the decay constant, and $t$ is time. To see a worked-out problem about radioactive materials in Moon rocks, take a look at www.tmt.ugal.ro/crios/Support/ANPT/Curs/deqn/a1/mrocks/mrocks.html.

# Studying Inductor-Resistor Circuits

If you have electrical circuits containing inductors and resistors, use the following differential equation to determine the current in the circuit:

$$L\ \frac{dI}{dt} + IR = V$$

where *I* is the current, *L* is the inductance, *R* is the resistance in the circuit, and *V* is the voltage as a function of time of the driving voltage source. To see how to handle this kind of problem, check out www.analyzemath.com/calculus/Differential_Equations, click the applications.html link, and scroll down to Application 5.

## Calculating the Motion of a Mass on a Spring

The motion of a mass on a spring that's moving horizontally (which means gravity isn't a factor) is given by this differential equation:

$$m\ \frac{d^2x}{dt^2} + kx = 0$$

where *x* is the location of the mass, *m* is the mass, and *k* is the constant of the spring. Visit hyperphysics.phy-astr.gsu.edu/hbase/shm2.html#c2 to see the solution.

# Index

## BUSINESS, CAREERS & PERSONAL FINANCE

**Accounting For Dummies, 4th Edition\***
978-0-470-24600-9

**Bookkeeping Workbook For Dummies†**
978-0-470-16983-4

**Commodities For Dummies**
978-0-470-04928-0

**Doing Business in China For Dummies**
978-0-470-04929-7

**E-Mail Marketing For Dummies**
978-0-470-19087-6

**Job Interviews For Dummies, 3rd Edition\*†**
978-0-470-17748-8

**Personal Finance Workbook For Dummies\*†**
978-0-470-09933-9

**Real Estate License Exams For Dummies**
978-0-7645-7623-2

**Six Sigma For Dummies**
978-0-7645-6798-8

**Small Business Kit For Dummies, 2nd Edition\*†**
978-0-7645-5984-6

**Telephone Sales For Dummies**
978-0-470-16836-3

## BUSINESS PRODUCTIVITY & MICROSOFT OFFICE

**Access 2007 For Dummies**
978-0-470-03649-5

**Excel 2007 For Dummies**
978-0-470-03737-9

**Office 2007 For Dummies**
978-0-470-00923-9

**Outlook 2007 For Dummies**
978-0-470-03830-7

**PowerPoint 2007 For Dummies**
978-0-470-04059-1

**Project 2007 For Dummies**
978-0-470-03651-8

**QuickBooks 2008 For Dummies**
978-0-470-18470-7

**Quicken 2008 For Dummies**
978-0-470-17473-9

**Salesforce.com For Dummies, 2nd Edition**
978-0-470-04893-1

**Word 2007 For Dummies**
978-0-470-03658-7

## EDUCATION, HISTORY, REFERENCE & TEST PREPARATION

**African American History For Dummies**
978-0-7645-5469-8

**Algebra For Dummies**
978-0-7645-5325-7

**Algebra Workbook For Dummies**
978-0-7645-8467-1

**Art History For Dummies**
978-0-470-09910-0

**ASVAB For Dummies, 2nd Edition**
978-0-470-10671-6

**British Military History For Dummies**
978-0-470-03213-8

**Calculus For Dummies**
978-0-7645-2498-1

**Canadian History For Dummies, 2nd Edition**
978-0-470-83656-9

**Geometry Workbook For Dummies**
978-0-471-79940-5

**The SAT I For Dummies, 6th Edition**
978-0-7645-7193-0

**Series 7 Exam For Dummies**
978-0-470-09932-2

**World History For Dummies**
978-0-7645-5242-7

## FOOD, HOME, GARDEN, HOBBIES & HOME

**Bridge For Dummies, 2nd Edition**
978-0-471-92426-5

**Coin Collecting For Dummies, 2nd Edition**
978-0-470-22275-1

**Cooking Basics For Dummies, 3rd Edition**
978-0-7645-7206-7

**Drawing For Dummies**
978-0-7645-5476-6

**Etiquette For Dummies, 2nd Edition**
978-0-470-10672-3

**Gardening Basics For Dummies\*†**
978-0-470-03749-2

**Knitting Patterns For Dummies**
978-0-470-04556-5

**Living Gluten-Free For Dummies†**
978-0-471-77383-2

**Painting Do-It-Yourself For Dummies**
978-0-470-17533-0

## HEALTH, SELF HELP, PARENTING & PETS

**Anger Management For Dummies**
978-0-470-03715-7

**Anxiety & Depression Workbook For Dummies**
978-0-7645-9793-0

**Dieting For Dummies, 2nd Edition**
978-0-7645-4149-0

**Dog Training For Dummies, 2nd Edition**
978-0-7645-8418-3

**Horseback Riding For Dummies**
978-0-470-09719-9

**Infertility For Dummies†**
978-0-470-11518-3

**Meditation For Dummies with CD-ROM, 2nd Edition**
978-0-471-77774-8

**Post-Traumatic Stress Disorder For Dummies**
978-0-470-04922-8

**Puppies For Dummies, 2nd Edition**
978-0-470-03717-1

**Thyroid For Dummies, 2nd Edition†**
978-0-471-78755-6

**Type 1 Diabetes For Dummies\*†**
978-0-470-17811-9

\* Separate Canadian edition also available
† Separate U.K. edition also available

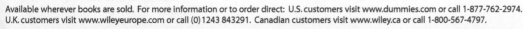
Available wherever books are sold. For more information or to order direct: U.S. customers visit www.dummies.com or call 1-877-762-2974.
U.K. customers visit www.wileyeurope.com or call (0) 1243 843291. Canadian customers visit www.wiley.ca or call 1-800-567-4797.

## INTERNET & DIGITAL MEDIA

**AdWords For Dummies**
978-0-470-15252-2

**Blogging For Dummies, 2nd Edition**
978-0-470-23017-6

**Digital Photography All-in-One
Desk Reference For Dummies, 3rd Edition**
978-0-470-03743-0

**Digital Photography For Dummies,
5th Edition**
978-0-7645-9802-9

**Digital SLR Cameras & Photography
For Dummies, 2nd Edition**
978-0-470-14927-0

**eBay Business All-in-One Desk Reference
For Dummies**
978-0-7645-8438-1

**eBay For Dummies, 5th Edition***
978-0-470-04529-9

**eBay Listings That Sell For Dummies**
978-0-471-78912-3

**Facebook For Dummies**
978-0-470-26273-3

**The Internet For Dummies, 11th Edition**
978-0-470-12174-0

**Investing Online For Dummies,
5th Edition**
978-0-7645-8456-5

**iPod & iTunes For Dummies, 5th Edition**
978-0-470-17474-6

**MySpace For Dummies**
978-0-470-09529-4

**Podcasting For Dummies**
978-0-471-74898-4

**Search Engine Optimization
For Dummies, 2nd Edition**
978-0-471-97998-2

**Second Life For Dummies**
978-0-470-18025-9

**Starting an eBay Business For Dummies
3rd Edition†**
978-0-470-14924-9

## GRAPHICS, DESIGN & WEB DEVELOPMENT

**Adobe Creative Suite 3 Design Premium
All-in-One Desk Reference For Dummies**
978-0-470-11724-8

**Adobe Web Suite CS3 All-in-One Desk
Reference For Dummies**
978-0-470-12099-6

**AutoCAD 2008 For Dummies**
978-0-470-11650-0

**Building a Web Site For Dummies,
3rd Edition**
978-0-470-14928-7

**Creating Web Pages All-in-One Desk
Reference For Dummies, 3rd Edition**
978-0-470-09629-1

**Creating Web Pages For Dummies,
8th Edition**
978-0-470-08030-6

**Dreamweaver CS3 For Dummies**
978-0-470-11490-2

**Flash CS3 For Dummies**
978-0-470-12100-9

**Google SketchUp For Dummies**
978-0-470-13744-4

**InDesign CS3 For Dummies**
978-0-470-11865-8

**Photoshop CS3 All-in-One
Desk Reference For Dummies**
978-0-470-11195-6

**Photoshop CS3 For Dummies**
978-0-470-11193-2

**Photoshop Elements 5 For Dummies**
978-0-470-09810-3

**SolidWorks For Dummies**
978-0-7645-9555-4

**Visio 2007 For Dummies**
978-0-470-08983-5

**Web Design For Dummies, 2nd Edition**
978-0-471-78117-2

**Web Sites Do-It-Yourself For Dummies**
978-0-470-16903-2

**Web Stores Do-It-Yourself For Dummies**
978-0-470-17443-2

## LANGUAGES, RELIGION & SPIRITUALITY

**Arabic For Dummies**
978-0-471-77270-5

**Chinese For Dummies, Audio Set**
978-0-470-12766-7

**French For Dummies**
978-0-7645-5193-2

**German For Dummies**
978-0-7645-5195-6

**Hebrew For Dummies**
978-0-7645-5489-6

**Ingles Para Dummies**
978-0-7645-5427-8

**Italian For Dummies, Audio Set**
978-0-470-09586-7

**Italian Verbs For Dummies**
978-0-471-77389-4

**Japanese For Dummies**
978-0-7645-5429-2

**Latin For Dummies**
978-0-7645-5431-5

**Portuguese For Dummies**
978-0-471-78738-9

**Russian For Dummies**
978-0-471-78001-4

**Spanish Phrases For Dummies**
978-0-7645-7204-3

**Spanish For Dummies**
978-0-7645-5194-9

**Spanish For Dummies, Audio Set**
978-0-470-09585-0

**The Bible For Dummies**
978-0-7645-5296-0

**Catholicism For Dummies**
978-0-7645-5391-2

**The Historical Jesus For Dummies**
978-0-470-16785-4

**Islam For Dummies**
978-0-7645-5503-9

**Spirituality For Dummies,
2nd Edition**
978-0-470-19142-2

## NETWORKING AND PROGRAMMING

**ASP.NET 3.5 For Dummies**
978-0-470-19592-5

**C# 2008 For Dummies**
978-0-470-19109-5

**Hacking For Dummies, 2nd Edition**
978-0-470-05235-8

**Home Networking For Dummies, 4th Edition**
978-0-470-11806-1

**Java For Dummies, 4th Edition**
978-0-470-08716-9

**Microsoft® SQL Server™ 2008 All-in-One
Desk Reference For Dummies**
978-0-470-17954-3

**Networking All-in-One Desk Reference
For Dummies, 2nd Edition**
978-0-7645-9939-2

**Networking For Dummies,
8th Edition**
978-0-470-05620-2

**SharePoint 2007 For Dummies**
978-0-470-09941-4

**Wireless Home Networking
For Dummies, 2nd Edition**
978-0-471-74940-0

# OPERATING SYSTEMS & COMPUTER BASICS

**Mac For Dummies, 5th Edition**
978-0-7645-8458-9

**Laptops For Dummies, 2nd Edition**
978-0-470-05432-1

**Linux For Dummies, 8th Edition**
978-0-470-11649-4

**MacBook For Dummies**
978-0-470-04859-7

**Mac OS X Leopard All-in-One
Desk Reference For Dummies**
978-0-470-05434-5

**Mac OS X Leopard For Dummies**
978-0-470-05433-8

**Macs For Dummies, 9th Edition**
978-0-470-04849-8

**PCs For Dummies, 11th Edition**
978-0-470-13728-4

**Windows® Home Server For Dummies**
978-0-470-18592-6

**Windows Server 2008 For Dummies**
978-0-470-18043-3

**Windows Vista All-in-One
Desk Reference For Dummies**
978-0-471-74941-7

**Windows Vista For Dummies**
978-0-471-75421-3

**Windows Vista Security For Dummies**
978-0-470-11805-4

# SPORTS, FITNESS & MUSIC

**Coaching Hockey For Dummies**
978-0-470-83685-9

**Coaching Soccer For Dummies**
978-0-471-77381-8

**Fitness For Dummies, 3rd Edition**
978-0-7645-7851-9

**Football For Dummies, 3rd Edition**
978-0-470-12536-6

**GarageBand For Dummies**
978-0-7645-7323-1

**Golf For Dummies, 3rd Edition**
978-0-471-76871-5

**Guitar For Dummies, 2nd Edition**
978-0-7645-9904-0

**Home Recording For Musicians
For Dummies, 2nd Edition**
978-0-7645-8884-6

**iPod & iTunes For Dummies,
5th Edition**
978-0-470-17474-6

**Music Theory For Dummies**
978-0-7645-7838-0

**Stretching For Dummies**
978-0-470-06741-3

# Get smart @ dummies.com®

- **Find a full list of Dummies titles**
- **Look into loads of FREE on-site articles**
- **Sign up for FREE eTips e-mailed to you weekly**
- **See what other products carry the Dummies name**
- **Shop directly from the Dummies bookstore**
- **Enter to win new prizes every month!**

\* Separate Canadian edition also available
\* Separate U.K. edition also available

Available wherever books are sold. For more information or to order direct: U.S. customers visit www.dummies.com or call 1-877-762-2974.
U.K. customers visit www.wileyeurope.com or call (0) 1243 843291. Canadian customers visit www.wiley.ca or call 1-800-567-4797.

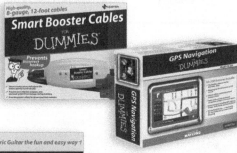